Refuting Creationism

Why Creationism Fails In Both Its Science Its Theology

Rosa Rubicondior

Refuting Creationism

Cover design: AI-generated image (ChatGPT4o)

Refuting Creationism

Third Party Copyright.

Third party copyright is acknowledged for work not currently in the public domain, reproduced here for criticism and analysis under intellectual rights fair use regulations.

In the opinion of the author, the minimum necessary for effective criticism and analysis whilst retaining the original context, has been reproduced in this work.

ISBN: 979-8345104989

© 2024 Rosa Rubicondior.
All rights reserved.

Refuting Creationism

As Kenneth Miller points out in his excellent book, Finding Darwin's God, for these claims to be true, God would have had to engage in massive subterfuge. For instance, since many of the observable stars and galaxies in the universe are more than ten thousand light-years away, a YEC perspective would demand that our ability to observe them could come about only if God had fashioned all of those photons to arrive here in a "just so" fashion, even though they represent wholly fictitious objects.

This image of God as a cosmic trickster seems to be the ultimate admission of defeat for the Creationist perspective. Would God as the great deceiver be an entity one would want to worship?

Is this consistent with everything else we know about God from the Bible, from the Moral Law, and from every other source—namely, that He is loving, logical, and consistent? Thus, by any reasonable standard, Young Earth Creationism has reached a point of intellectual bankruptcy, both in its science and in its theology. Its persistence is thus one of the great puzzles and great tragedies of our time. By attacking the fundamentals of virtually every branch of science, it widens the chasm between the scientific and spiritual worldviews, just at a time where a pathway toward harmony is desperately needed. By sending a message to young people that science is dangerous, and that pursuing science may well mean rejecting religious faith, Young Earth Creationism may be depriving science of some of its most promising future talents.

Kenneth Miller, *The Language of God* pp 176-177

Contents

Introduction 9
    Teleological thinking causes creationism. 10
    Misinterpretation of natural phenomena: 11
    Attribution of Intentionality 11
    Reinforcement of unscientific prejudice: 12
    Confirmation bias: 13
    Essentalism: 13
    Morton's Demon: 14
    Soft tissue in T. rex fossil fallacy. 15
    What Genesis really says. 17
    The Galaxy Song. 18

Creationism's Fatal Flaw 21
    Throwing Stones at Science 21
    How and why science works 21
    Occam's Razor 21
    Multiplying entities needlessly 24
    Confirmation bias 25
    God of the Gaps. 26

Religious Fundamentalism Dressed in a Lab Coat 27
    The Wedge Strategy 28
    Aims and Objectives 28
    Strategy 28
    Tactics 29
    Resulting court cases 30
        Epperson v. Arkansas (1968) 30

  Edwards v. Aguillard (1987) 30
  Peloza v. Capistrano Unified School District (1994) 30
  Freiler v. Tangipahoa Parish Board of Education (1999)
  31
  Selman v. Cobb County School District (2005) 31
  Kitzmiller v. Dover Area School District (2005) 31
  Association of Christian Schools International v. Stearns (2008) 32
 Failure of The Wedge Strategy. 32
 What's Wrong With Intelligent Design? 33
  Genetic Entropy. 36
  Devolution 37
How Do We Know How Old Earth Is? 39
 Uranium-Lead (U-Pb) Dating. 41
 Potassium-Argon (K-Ar) Dating: 43
 Rubidium-Strontium (Rb-Sr) Dating: 43
 The fallacy of changing radioactive decay rates: 43
 Luminescence Dating: 45
 Paleomagnetic Dating: 46
 Biostratigraphy: 47
 Carbon Dating. 47
Evolution Works and Why Creationists Think It Doesn't 53
 Darwinian Natural Selection 55
  Brassica fruticulose 57
 Genetic Drift 58
  The Amish and Ellis van Creveld syndrome. 61
  Northern elephant seal 61

Contents

    Cheetahs 61
    Laboratory fruit flies 62
Hybridization. 62
    Homo sapiens and Neanderthals 63
    Polyploidy As a Cause of Speciation 64
    Edible Frogs 65
    The Marbled Crayfish 66
    New Mexico Whip-Tailed Lizard. 67
Horizontal Gene Transfer 68
    New Genetic Information and the Second Law of Thermodynamics. 68
    A Fruit fly/bacteria hybrid 70
    Photosynthesising Green Sea Slugs. 71
    Endosymbiosis 71
    Parasitic plants 71
    New Metabolic Pathways in Fungi. 72
    The Asian Longhorned Beetle 72
    Bdelloid Rotifers 73
Species, Hybrids and Kinds 75
    The Biblical Kind and Why Biologists Don't Use it. 75
    Hybridization and Scientific Taxonomy. 77
        Carrion Crow/Hooded Crow 78
        Red Shepherd's Purse 84
        Welsh Ragwort 86
        Evolution of a New Species Recorded in Stone Age Cave Art 87
    Ring species 89

Larus gulls 89
The Greenish Warbler 90
Ensatina salamanders of California. 90
A Human ring species? 91
Barriers to hybridization 92
Pre-zygotic barriers 92
Post-zygotic barriers 93
European finches 94
When pre-zygotic barrier fail – hybrid geese and ducks.
.... 96

Macro-Evolution Vs Micro-Evolution 101
Definitions and Applications 101
Micro-evolution 101
Macro-evolution 102
Key Conceptual Differences 102
Scientific Definition of Macro-evolution 103
The Controversy in Historical Context 103
Summary: 103
The Wall Lizards of Pod Mrcaru 105
Rapid Evolution in agricultural weeds 107
Barnyardgrass 107
Waterhemp 107
The Golden jellyfish 108
The East African Cichlids 111
London Underground Mosquito 113
Heliconius Butterflies 115
Apple Maggot Flies (*Rhagoletis pomonella*) 115

Contents

New Stickleback Species 116
Transitional Forms 117
   Evolution of the Giant Pterosaurs 118
   The First Arthropods 119
   Tiktaalik roseae 120
      Tiktaalik roseae and the Terrestrial Vertebrate Jaw 121
   Transition to The Mammalian Middle Ear. 122
   Transitioning Into Insects 123
   A Lagerstätte of Transitional Fossils 125
   A Feathered Pterosaur 126
   Transitional Crabs 127
   The Panda's Transitional 'Thumb' 129
   The Spider's Tail 130
   Transitional Hominins 131
      Australopithecus sediba 131
      Sahelanthropus tchadensis 133
      Australopithecus afarensis (Lucy) 135
      Homo naledi 136
      Evolution of the Balance Organ for Bipedalism 137
   Transitional Forms and Creationist Duplicity. 139
Something From Nothing 141
   The Big Bang. 141
   Abiogenesis 143
   The Genetic Code 147
      The Error Minimization Hypothesis, ... 149
      The Stereochemical Hypothesis 149
      The Co-Evolution Theory 149

- The Frozen Accident Theory 150
- How Creationists Mislead Their Supporters 151
  - Radio-halos 151
  - False Witnessing and Misrepresentation. 153
  - The 'Progression of Man' Fallacy 158
  - 'Lucy', ... 159
  - Heidelberg Man 159
  - Nebraska Man. 160
  - Piltdown Man 163
  - Peking Man 168
  - Neanderthal man 169
  - New Guinea Man 170
  - Cro Magnon Man 170
  - Modern Man 171
- Quote Mining 172
  - Lack of Identifiable Phylogeny 172
  - Karl Popper 176
  - Charles Darwin 178
  - Deliberate misrepresentation of science 182
- The Fine-Tuned Fallacy 187
  - The Anthropic Principle 187
  - A Misuse of statistics 187
    - The Vanishingly Small Probability Tactic 189
    - The Law of Large Numbers: 189
    - The Central Limit Theorem (CLT): 190
    - Reduced Margin of Error: 190
    - Guesses and Gobbledygook. 191

Contents

Why The Bible Cannot Be Taken Seriously.........................193

  Noah and The Genocidal Flood...193

    The changing definition of 'evolution'........................193

    A whole lot of Water 195

    Was it a boat or a box?...195

    Every living substance destroyed! 196

    What was floating in the water? 197

    Where is the sediment?...197

    What about fresh air?...197

    Who played host to the parasites and STDs?..............198

    A dove finds a living tree. Oops! 198

    What did the survivors eat? 198

    Nothing could survive on a sterile anoxic Earth..........199

    Just too implausible to be taken seriously. 199

    Why was the tale so ludicrously implausible?............200

  The Tall Tale of The Tower of Babel. 200

    How many generations since Noah?.............................201

    How many languages? 201

    An Omniscient God Gets a Surprise............................203

  Lot of Nonsense 204

    God get a lesson in Morality..204

    Lot offers his two virgin daughters to a mob to be gang raped..205

    Everyone is blinded – but not everyone.......................205

    Lot's wife achieves fame in her own right. 206

    Lot gets drunk – but not THAT drunk. 206

    What it was really all about. 206

The Fishy Tale of Jonah..207
    Jonah is given a job to do..208
    Jonah tried to flee to Tarshish...208
    Sailors believe Yahweh is out to get Jonah. 208
    A Home from home in the Stomach of a fish. 208
    The transgender fish gets sick...209
    God tries again..209
    Jonah warns the people of Ninevah 210
    Has the king taken leave of his senses? 210
    Jonah throws a tantrum because Nineveh is not destroyed!..211
    The Magic gourd..211
Would You Adam and Eve It?...212
    The magical creation of Adam..212
    The cloning of Eve..213
    The Gnostics 213
    Forbidden fruits and God's lie to Adam 214
    The serpent tells Eve the truth. 214
    The divine set-up. 215
    Men and women are allotted their roles in life. That'll teach 'em! 216
    Have you guessed who wrote it yet? 216
Conclusion 217
References..221
Index 235
Other Books by Rosa Rubicondior 245

Introduction

Creationism is religious fundamentalism that believes a supernatural deity with creative powers was personally responsible for all of existence; that without a creator god there would be nothing; a structureless void.

It comes in several different forms from believing in a creator god who micro-manages the entire universe, without whom there would be no physical laws, no physics or chemistry, not even atoms with orbital electrons, to Deism which believes a supernatural entity or force created the entire Universe (or even the multiverse) as a set of initial conditions, lit the blue touch-paper and then has played no further part in the process.

Any debate with the latter is over whether the initial conditions were set to produce some pre-planned outcome or whether the creator had no idea how the Universe would progress once it was set in motion and is now merely a non-interventionist observer.

The former view tends to be popular with those who want there to be some purpose to it all, and especially if the purpose was to produce humans in general and them in particular, so it becomes a scientifically-aware form of anthropocentrism where all the forces of nature are arranged so they produce us, here today with everything as God intended.

All forms of creationism have humans as a special case and the reason for the Universe's existence is as somewhere for humans to exist, so, in reality, creationism is a form of narcissistic anthropocentrism, a view for which not a single scrap of justification exists.

Objectively, the whole of nature could with equal justification be used to justify the special nature of anything, animate or inanimate, from elephants to galaxies and black holes, of which there are probably more than there are or have ever been humans in the entire history of humanity.

It is, of course, far more logical to assume that nothing is so special as to be the reason for the Universe's existence; that everything is simply the result of fundamental forces acting over time.

Creationism is, at its heart, the product of teleological thinking, i.e., the belief that an agency of some sort must be behind everything because nothing can happen unless something wants it to or makes it. In other words, the belief that everything is sentient and has cognitive abilities like humans.

Teleological thinking causes creationism.

Teleological thinking is a mode of thinking that children use to explain how things work, as was shown in a research paper by four French psychologist, Wagner-Egger, *et al.* (1).

In that paper they defined it as "…the attribution of purpose and a final cause to natural events and entities" and explained that "Although teleological thinking has long been banned from scientific reasoning, it persists in childhood cognition, as well as in adult intuitions and beliefs. They showed that teleological thinking when retained into adulthood is the cause not only of creationism but also of conspiracism and other anti-scientific views with:

> Because teleological and animist thinking are part of children's earliest intuitions about the world and are resilient in adulthood (2), (3) they thus could be causally involved in the acquisition of creationist and conspiracist beliefs. However, our results do not rule out the possibility that acceptance of such beliefs could, conversely, favour a teleological bias. Yet, in both cases, the 'everything happens for a reason' or 'it was meant to be' intuition at the heart of teleological thinking not only remains an obstacle to the acceptance

of evolutionary theory, but could also be a more general gateway to the acceptance of anti-scientific views and conspiracy theories.

Teleological thinking, although an immature form of reasoning, is seductive especially in those prone to seek easy to understand answers to complex problems.

For example, it is easier to assume that something sentient makes a chemical reaction happen; that the atoms and molecules involved are in some way conscious of those desires or instructions or of the laws made for them to follow, rather than learn the complex energy changes and interactions of electrostatic forces that produce particular chemical bonds, especially at the level of quantum mechanics (which can sound like magic with its dualities and 'spooky actions at a distance').

The errors that teleological thinking can lead to include:

Misinterpretation of natural phenomena:

- For example, assuming purpose behind essentially random events then looking for the cause of natural disasters in terms of divine punishment (with someone to blame) or assuming evolution has a purpose – to create humans, to create a better organism or that so an animal or plant can live in a particular environment.

- One common form of anti-evolution 'killer' question is "Why did a chimpanzee decide to have a human baby?" or "If some chimpanzees decided to become human, why didn't the others?"

Attribution of Intentionality:

- For example, birds must have decided to fly, or they wouldn't have evolved wings (so how did they know they needed wings in order to fly?). This can lead to subtle biases even in people who accept evolution, by assuming evolution is goal-orientated or purpose-driven, e.g., a virus evolves into new variants 'so that' it can evade the antibodies vaccines produce, as though the virus is conscious of the threat it faces from antibodies and works out how to change to evade them.

- A great deal of discussion of evolution involves casual and possibly unconscious reference to purpose as a driver of evolution. For example,:

    > The tiger has stripes so it can hide from its prey as though a tiger has conscious control over its fur and is aware of the perceptions of its prey species.

    > Rattlesnakes have a rattle so they can warn us not to tread on them, as though the rattlesnake is aware of the dangers and decides to make the dead skin at the end of its tail into a rattle.

    The problem with that is it plays to the prejudice of those looking for reasons to dismiss evolution because it makes manifestly absurd assumption which the science-denier can present as evidence of the stupidity of scientists and so to the 'superior science' of creationism.

Reinforcement of unscientific prejudice:

- For example, the belief that viruses and bacteria, or even cancers and other illnesses, exist for a purpose – to punish us or to make us ill for evil purposes.

- Mental health still carries a stigma from the days when it was assumed to be caused by 'moral weakness' in allowing evil demons to possess the victim. By the same erroneous thinking 'moral weaknesses', or sexual promiscuity, was once classified as a mental illness. The notion of possession is still promulgated by, for example, the Catholic Church which still appoints 'exorcists' and by fundamentalist 'faith healers' who 'drive out evil' in the 'name of Jesus' from the body of the sick – a superstition given credence by the story of Jesus driving out demons and sending them into a herd of pigs who promptly committed suicide by jumping over a cliff.

- Belief in witchcraft and all the horrors that led to was driven by the belief that witches, when possesses by demons, could cause evil things to happen. In other words, the belief that unfortunate events had a teleological cause in the shape of evil agents taking human form. Intent and agency to justify unscientific prejudice.

Confirmation bias:

- A teleological explanation can seem so obvious to some people that any contrary evidence can set up cognitive dissonance where the person then rejects anything that contradicts the bias.

Essentalism:

- A teleological thinker can easily fall into the trap of assuming purpose or intent to processes or living things. One powerful reason for rejecting the 'materialist', natural explanation for observable phenomena is that it denies a 'purpose' for life. The assumption is that a person is born for a purpose – to fulfill some assumed plan; that their life is for something, without realising that they are free to give themselves whatever purpose they wish.

- A creationist will argue that every life must have a purpose therefore something must have determined that purpose – and of course has a divine plan to ensure the purpose is fulfilled. A persuasive argument for people whose life would otherwise appear hopeless or unimportant in the 'scheme of things', hence the appeal of religion to those in the lower strata of stratified societies.

And that leads naturally to the cognitive device that many if not all creationists employ to retain their cognitive bias by shutting out any arguments and evidence against it – the so-called Morton's Demon.

Morton's Demon:

Glenn Moreton is a former Young Earth Creationist who, through a study of geology, came to realise Earth was much older than the 6-10,000 years that most YECs believe it to be.

He realised he had been using a psychological devise to maintain his creationism in a way similar to 'Maxwell's Demon'. Maxwell's Deamon was a thought experiment by the physicist James Clark Maxwell, who suggested a demon could stand in a doorway connecting two rooms and allow only fast moving (hot) particles to pass one way and only slow moving (cold) particles to pass the other way. This would create a temperature gradient that could be used to perform work. Since the demon would not expend energy, this appeared to violate the Laws of Thermodynamic and create a perpetual motion machine.

This has since been refuted because the demon would need to expend energy in observing the speed of approaching particles and stop those heading in the 'wrong' direction, but Glenn Morton realised he had used an analogous demon as a filter to maintain his Young Earth Creationism by only allowing things

Introduction

that reinforced his beliefs to pass though his door of perception while stopping anything that would threaten it.

Once he removed this 'demon' he was flooded with evidence for an old Earth and against the notion of a young earth.

You can see Morton's Demon at work in any online debate group where a creationist will frequently ask for evidence for something or other, and having been given that evidence in the form of an article or research paper, will to reject it out of hand, or not even read it, and ask for the same evidence for the same thing a few days or weeks later, or in a different group.

Soft tissue in T. rex fossil fallacy.

Similarly, a false claim will be presented as evidence for a young earth or against evolution, such as the false claim that the palaeontologist Professor Mary Higby Schweitzer found soft tissues and DNA in a Tyrannosaurus rex fossil, "proving it was just a few thousand years old" (4) (5), and even the egregious lie that she 'carbon-dated' it to a few thousand years old. The claim will be refuted with reference to a denial by Professor Schweitzer herself (6) and the fact that she co-authored a research paper explaining how soft tissues such as collagen can be preserved for many millions of years in the presence of iron (7), and yet the same claim will be made again by the same YEC in a different group.

Professor Schweitzer represents an example of how creationists can hold two mutually contradictory views simultaneously and how, as Wagner-Egger, *et al.* showed (1), creationism and conspiracism are both caused by the same teleological thinking retained from infancy.

On the one hand, she is a brilliant scientist who has 'proved' that dinosaurs lived just a few thousand years ago, and, when she refutes and rejects that claim, she is part of the conspiracy of evil trying to turn creationists away from God. The guardian demon is allowing through the "dinosaur soft tissue proves a young earth" false claim while shutting out the scientific evidence that this is not so and protecting the creationists against the forces of evil, because anyone who provides evidence against a young earth must be part of a vast scientific Satanic conspiracy of evil.

Science provides indisputable 'proof' when it appears to support creationism but can be dismissed as vast conspiracy when it does just the opposite.

Evidence of Morton's Demon at work can be seen in professional creationist apologists who will take part in a public debate one evening in which they will make particular claims only to have them systematically refuted by their opponent in the debate, and then repeat exactly the same arguments in another debate or on television a day or two later, as though they have never heard any counter-arguments and are still convinced of the truth of their arguments.

The alternative view, which undoubtedly applies to many professional creationists who earn their living providing creationists with spurious arguments and confirmation is that they are not unaware of a psychological filter working to reduce the cognitive dissonance unwanted facts will cause, but are fully aware that their arguments are fallacious and intelligently designed to mislead their audience. They are confident that the audience will never fact-check any of the claims provided they tell them what they want to hear and will reject counter-arguments because they at least will have come armed with a guardian demon keeping them safe from the machinations of Satan.

In the following chapters I will explore the various arguments that are traditionally employed by creationists in their attempt to dispute science and reject the scientific evidence that Earth is billions of years old in a vast Universe that is several times older, and that living organisms are the result of a natural evolutionary process sharing a last common ancestor (LUCA)

Introduction

that arose by chemical and physical processes early in the history of the planet.

An integral part of fundamentalist creationism is bible literalism which holds that the Bible, and especially the first chapter, Genesis, is a literal account of real history written or inspired by the creator god described in it, therefore the account of a global genocidal flood, and the Tower of Babel are real history, as are later accounts of the sun standing still in the sky while the moon hid in a valley.

What Genesis really says.

It's worth bearing in mind that a literal reading of the description of the Universe in Genesis is a description of a small flat planet with a dome, or firmament, over it to keep the water above the sky out (Genesis 1:6-10). The sun, moon and stars are stuck to the dome (Genesis 1:16-17) and the stars can be shaken loose by earthquakes when they fall to earth (Mark 13: 25) and can even be trampled on by goats (Daniel 8:10).

Nothing in that universe existed outside a day or two's walk of the small area of the Middle East where the Hebrew origin myths were part of the local folklore in the Bronze Age.

Much of that, which until a few hundred years ago was considered indisputable fact as reveal by the creator, is now regarded, even by most bible-literalist creationists, as metaphorical or allegorical. This represents a considerable retreat by Christianity in the face of scientific evidence that so much of the Bible is factually incorrect.

Earth did not appear fully formed in its present state but coalesced along with the other planets in an accretion disk around a young, second or third generation star and is the result of geological processes over billions of years including plate tectonics and volcanism causing climate change and rises and falls in sea level.

Species did not suddenly arise fully formed in their present state without ancestry but are the result of evolutionary processes over many millions of years as earlier species diverged in response to local environmental changes, ultimately driven by the geological forces shaping the planet.

And Earth is not the center of the universe occupying a fixed position while the sun, moon and stars orbit around it but is a planet orbiting one of half a trillion stars in the outer arm of one of half a trillion galaxies in a vast expanding Universe.

In the words of The Galaxy Song' by Eric Idle from 'Monty Python's The Meaning of Life;

The Galaxy Song.

Just remember that you're standing on a planet that's evolving
And revolving at 900 miles an hour.
It's orbiting at 19 miles a second, so it's reckoned,
The sun that is the source of all our power.
Now the sun, and you and me, and all the stars that we can see,
Are moving at a million miles a day,
In the outer spiral arm, at 40, 000 miles an hour,
Of a galaxy we call the Milky Way.

Our galaxy itself contains a hundred billion stars;
It's a hundred thousand light-years side to side;
It bulges in the middle sixteen thousand light-years thick,
But out by us it's just three thousand light-years wide.
We're thirty thousand light-years from Galactic Central Point,
We go 'round every two hundred million years;
And our galaxy itself is one of millions of billions
In this amazing and expanding universe.

Our universe itself keeps on expanding and expanding,
In all of the directions it can whiz;
As fast as it can go, at the speed of light, you know,
Twelve million miles a minute and that's the fastest speed there is.

Introduction

So remember, when you're feeling very small and insecure,
How amazingly unlikely is your birth;
And pray that there's intelligent life somewhere out in space,
'Cause there's bugger all down here on Earth!

    (Galaxy Song lyrics by Eric Idle/John du Prez
    © Universal Music Publishing Group) (8)

Lastly, as I always say in my books, do not take anything I say for granted. Always check. I have provided references for any substantive claims I have made. Please use them to fact check anything I have said.

Refuting Creationism

Creationism's Fatal Flaw

It quickly becomes apparent from reading any creationists literature that the entire cult depends on the childishly absurd, and breathtakingly arrogant belief that if you can prove your opponent wrong, you win the argument by default, so don't need to produce any evidence to support your claims.

Throwing Stones at Science

It manifests itself in constantly attacking science and throwing stones at straw man parodies of it whilst never providing a single scientific datum supporting the claim that everything was magically created out of nothing by an omnipotent creator that somehow exists outside space and time and yet is able to influence how chemistry and physics works without being detectable.

But that's not how science works, and why creationism isn't the science its proponents claim.

How and why science works

Science works by producing evidence that supports a hypothesis and by failing to falsifies it.

For example, Germ Theory became an accepted scientific theory for the cause of certain diseases because scientists firstly showed that germs existed, then showed they could cause infections by recovering them from an infected patient and using them to cause the same disease in another subject. At every stage in that argument, every entity was proved to exist and do what advocates of the theory claimed they do.

Occam's Razor

The underpinning philosophy of this approach to science was formulated by no other than an English Franciscan friar, William of Ockham or Occam (c. 1278-1347) who reasoned that the most vicarious answer to a problem was most probably the correct one, so he formulated what became known as Occam's Razor – a philosophical device for removing any element in a line of reasoning that isn't required, or which adds nothing but additional complexity to the argument.

In brief Occam's Razor says:

"Do not multiply entities needlessly."

In other words, add nothing to your argument that you can't show exists and is germane to the argument.

The fatal flaw in creationism is that it does exactly that, for no other reason than to make the advocate feel more important than he or she would otherwise feel, or to satisfy a desire for there to be a god.

And yet every explanation that science has ever produced to explain anything previously not understood has never required a god.

I will illustrate this with a simple example – the petrol-powered internal combustion engine;

A petrol-powered internal combustion engine works by a fuel (petrol) being mixed with air which contains oxygen, being mixed in the carburetor to give the right proportions. This is then passed into the combustion chamber and compressed when the piston moves up.

At the right time, an electrical spark ignites the mixture which cause the oxygen and the petrol to undergo a chemical reaction to produce carbon dioxide and water vapour which occupies a larger space than the original mixture, so this pushes the piston down in the chamber providing power to propel the vehicle through a system of gears.

The piston then moves up in the cylinder and the carbon dioxide and water vapour are expelled to clear the chamber for the next cycle. Of course, there is a mechanical system of valves to open and close to allow the fuel mixture in and the exhaust mixture out, but that is the simple explanation of the combustion/exhaust cycle of an internal combustion system.

We could add more in the form of an explanation of the chemical reaction between the large molecules of petrol and the oxygen, to give lots of small molecules of carbon dioxide and water gas by adding details of chemical bonds, orbital electrons and electron rearrangement and quickly get into quantum mechanics. All of which is explained by basic laws of thermodynamics.

We could go into how the physics of a gas means this has a greater volume than the original mixture because there are many more molecules in it and get into an explanation of the Gas Laws and Avogadro's Number and we have a complete explanation of how a petrol-powered internal combustion engine works.

And every entity in that explanation has *a priori* been established with empirical evidence before we included it in the explanation. In other words, our explanation is based on established science.

Then, we can begin to add more complexity to that explanation simply to put a god into the answer. For instance, as I was once assured by a fundamentalist creationist when discussing how a bullet works, a god (of course the locally popular one) is required to tell the chemicals how and when to react and what to produce and then to tell the products of combustion to occupy more space than the fuel/air mixture. To a teleological mind, everything is sentient and needs to be told what to do and how and when to do it.

So, we have now needlessly multiplied entities simply to satisfy a need for a god and for simplistic answers to explain physics and chemistry to a teleological thinker who never progressed beyond toddlerhood in his thinking ability.

So now we need to add yet more entities to explain how a god makes things work and how it came to exist and acquire those powers, and why the proposer selected that particular god. And we end up with a vastly more complex answer to the original question (how does a petrol-powered internal combustion work?) when we had a more vicarious answer in which no gods were required and to which adding one simply complicated the answer without adding anything to it.

And yet that's what creationists insist on doing all the time and even believe they have a better answer, while never providing any empirical *a priori* evidence that their additional entities actually exist or could do the things they are claimed to do.

Multiplying entities needlessly

In fact, even considering a god as part of the explanation would first require the scientist to demonstrate the truth of two propositions (adding more complexity and multiplying entities needlessly). They would need to:

1. Explain the origins of a supernatural entity with the powers to do whatever the argument requires it to do, and that would have to include how such an entity self-assembled from nothing, complete with all the complexity and information to perform whatever function is required by the hypothesis.

   Simply waving that requirement aside with the assertion that it has special properties that don't need evidence because it has always existed (special pleading is simply cowardly avoidance and doesn't explain how the advocate has the right to grant their 'special entity' special exemptions not granted to the rest of scientific evidence).

2. Demonstrate authenticated instances of that entity ever doing what it is claimed to have the powers to do.

No creationist advocate has ever been able to demonstrate those two things, so there is no justification for including their god in the explanation for anything. Not only is it never demonstrated but the excuse for including it anyway add yet more levels of complexity and more (unexplained) entities to the argument:

- How does a god make chemistry and physics do things they can't do without it?

- What is the god made from and how does it acquire information?

- How can an omniscient, inerrant god, who has always known in advance what it will do, change its mind or make any decisions or is it constrained in a predetermined universe by its own omniscience (in other words, is it in fact powerless)?

So, including a god in any explanation simply multiplies entities needlessly and produces an infinitely complex 'answer' that raises more questions that it can ever answer.

It should be pared away, using Occam's Razor.

Confirmation bias

That it satisfies their ignorant incredulity or satisfies some superstition that they acquired from their parents or authority figures in their culture, is not a reason to insert it into any argument that is complete without it. This is exactly the personal bias that the scientific method is designed to exclude.

One manifestation of this intellectually dishonest approach to scientific questions is the so-called 'Intelligent Design' argument, where the entire argument invariably boils down to the argument that science can't (or hasn't yet – which equates to the same thing in creationists circles) explained how something evolved naturally, therefore a god must have designed and made it.

Not only is that a blatant non sequitur, but it contravenes Occam's Razor to include it.

God of the Gaps.

It's the classic God of the Gaps argument that seeks to force fit creationism's god into any gaps, real, imaginary or manufactured, in science, in the forlorn and touchingly optimistic belief that, whereas every time science has closed such a gap, no gods were found to be needed, yet this time, there creationism's god will be found in all its glory and creationists will be triumphant.

So, until a creationist can provide the necessary explanation for the origins of god and the authenticated evidence of its power to create, they don't have an argument worth considering.

As Christopher Hitchins said,

"That which can be asserted without evidence can be dismissed without evidence."

Religious Fundamentalism Dressed in a Lab Coat

Creationism is religious fundamentalism that has a problem with the American Constitution and the law as it stands in many other countries where there is a prohibition on teaching religions in schools or at least, as in the UK from teaching creationism in a science class in schools.

The major obstacle is the clause in the US Constitution that was intended to maintain a 'Wall of Separation' between church and state, as advocated by the Founding Fathers, notably Thomas Jefferson, whom, along with others such as Geoge Washington and Thomas Payne were Deist Humanists who believed in religious freedom and tolerance.

The clause, known as the 'Establishment Clause' is in the first clause in the First Amendment:

> Congress shall make no law respecting an establishment of religion or prohibiting the free exercise thereof; or abridging the freedom of speech, or of the press; or the right of the people peaceably to assemble, and to petition the Government for a redress of grievances.

Which has been interpreted by the Supreme Court as applying at all levels of government and governmental institutions such as public-school boards which manage publicly funded schools.

It was against this legal backdrop that the 'Discovery Institute' was established to try to win political and public support for overturning that clause and allowing creationism to be taught in public schools as a step on the road to creating a fundamentalist Christian theocracy in the USA. To that end, the Discovery Institute produced the secret 'Wedge Strategy' in which teaching 'Intelligent Design' as science was to be the thin end of the wedge, opening up a gap between the law and public opinion.

The Wedge Strategy

The Discovery Institute's "Wedge Document," (9) also known as the "Wedge Strategy," was drawn up in the late 1990s and outlined a 5-year plan to promote intelligent design (ID) and challenge the teaching of evolution in the U.S. Originally intended as a secret, the document was leaked in 1999, and was at first disowned by the Discovery Institute, but later accepted by them as genuine (10).

Aims and Objectives

It revealed subversive goals for the Intelligent Design movement. The aim was to reshape public perceptions about evolution and establish intelligent design as an alternative science. The goal was (and still is) to replace what it describes as "materialistic" science with a science that conformed with Christian beliefs. The document identifies a "wedge" approach, aiming to destroy the scientific consensus on evolution by emphasizing intelligent design.

The *raison d'etre* of the Discovery Institute and its Wedge Strategy is recognition of just how much science undermines the Bible and the basis of Christianity, with its unelected and unaccountable hierarchies, social control and political influence.

Strategy

The strategy had three phases:

1. Scientific Research, Writing, and Publication: This phase includes producing and publishing ID-friendly 'scientific' papers, books, and articles, with the objective of making ID look like real science, so garnering respect within the scientific and academic communities.

2. Publicity and Opinion-Making: The second phase aims to shape public opinion through media and public relations. In other words, a PR campaign to misinform

the public and encourage them to question evolution and view ID as a legitimate science.

3. 'Cultural Renewal': The final phase is to influence educational, legal, and cultural institutions (in other words, to insert Christian fundamentalism into all aspects of American cultural and political life). This includes campaigning for ID to be taught in schools with the aim of altering how science is perceived and taught, especially in public education.

Tactics

The strategies and tactics were (and still are):

- Building alliances with conservative and religious groups.
- Presenting evolution as a flawed, controversial theory to produce doubt in the minds of the public with ID as a better alternative.
- Framing ID in secular terms to hide the fact that it is religion in disguise.

However, the whole tenor of the document reveals its theistic agenda and strong alignment with the aims and objectives of religions, betraying the fact that the ID movement is primarily a religious one aiming to subvert the secular constitution of the USA and replace it with a fundamentalist theocracy.

In their 2019 response to the revelation of their Wedge Document, the Discovery Institute list the 'Wedge Strategy Progress Summary' (10) (pp 17-18) which contains a list of books and other publications which should be read with the aims and objectives of the Institute in mind. Sadly, the pdf document is copy protected which precludes it being reproduced in this book. However, the list of books includes published or pending books by William Dembski, Paul Nelson, and Stephen Myer, and of course Michael J. Behe's *Darwin's Black Box*. (11).

So transparent is this subversive objective that the Wedge Strategy has been cited in court as evidence that attempts to mandate teaching ID as science in public schools is an attempt to circumvent the Establishment clause.

Resulting court cases

The result of the Wedge Strategy has been a series of court cases challenging states and boards of education attempts to implement part of the strategy and insert ID into science classes. The most notable such cases are:

Epperson v. Arkansas (1968)

This Supreme Court case struck down an Arkansas law that banned the teaching of evolution in public schools. SCOTUS ruled that the law violated the Establishment Clause of the First Amendment by promoting a particular religious viewpoint. The decision set a precedent that teaching evolution could not be prohibited simply because it conflicted with religious beliefs.

Edwards v. Aguillard (1987)

SCOTUS ruled that , Louisiana's "Creationism Act", which required that if evolution was taught, "creation science" also had to be presented as an alternative. SCOTUS struck down the law as unconstitutional, stating it was intended to promote a particular religious doctrine. This case reinforced that public schools could not mandate teaching creationism alongside evolution.

Peloza v. Capistrano Unified School District (1994)

John Peloza, a biology teacher, claimed that teaching evolution was a form of religious indoctrination that violated his religious freedom. The Ninth Circuit Court of Appeals ruled against him, deciding that evolution is a scientific theory, not a religious belief, and thus did not infringe upon his rights.

Freiler v. Tangipahoa Parish Board of Education (1999)

A Louisiana school board required teachers to read a disclaimer before teaching evolution, stating that it was a "theory, not fact" and encouraging students to seek alternatives. The Fifth Circuit Court of Appeals found the disclaimer unconstitutional because it endorsed a particular religious view.

Selman v. Cobb County School District (2005)

Cobb County in Georgia placed stickers in biology textbooks stating that evolution was "a theory, not a fact." The district court ruled that the stickers were unconstitutional because they promoted religious beliefs, effectively aligning with creationist views. Although the case was later settled without a final judgment, the school board removed the stickers.

Kitzmiller v. Dover Area School District (2005)

The Dover, Pennsylvania, school board mandated that students be introduced to "intelligent design" as an alternative to evolution. Federal Judge John E. Jones III ruled that ID was a religious view, not science, and its inclusion in the science curriculum violated the Establishment Clause. This was a major defeat for the ID movement, as the ruling was comprehensive and based on a detailed analysis of science and religion. It was during this trial the Michael J. Behe (witness for the defendants (Dover District School Board)) was forced to agree under cross examination by Mr Eric Rothschild, (for the plaintiff):

> ...there are no peer reviewed articles by anyone advocating for intelligent design supported by pertinent experiments or calculations which provide detailed rigorous accounts of how intelligent design of any biological system occurred.

In his summary, Judge John Jones III remarked on how fundamentalist Christians had lied under oath. In other words, they lied under oath to try to trick a judge into allowing them to lie to children at public expense.

> When you show the world you know you need to lie for your faith, you show the world you know your faith is for fools who'll believe lies

Association of Christian Schools International v. Stearns (2008)

The Association of Christian Schools International sued the University of California system, arguing that the university's refusal to accredit certain Christian school courses that used creationist textbooks was discriminatory. The court ruled in favor of the university, finding that the textbooks did not adequately prepare students in science, as they taught religious beliefs instead of scientific principles.

Failure of The Wedge Strategy.

There are no court cases in the USA where the view the Wedge Strategy seeks to insert into public perception - of ID as a real science - has prevailed.

It is now 34 years since the Discovery Institute launched its 5-years strategy to subvert the US Constitution by fooling the public, educators and legislators into thinking ID is real science, as the thin end of its ultimate end of a theocratic USA. In that time, support for the creationist view that God created humans in their present form in US public opinion has declined from 45% to 37% while acceptance of the idea that humans evolved over time has increased from 37% to 55% and 24% believe God was not involved in the evolution of humans. Evolution is now widely accepted by Generation Z and most people with a college education or higher.

A Gallop Poll in 2024 concluded:

> As Americans have become less religious over the past four decades, their beliefs about the origin of humans have shifted, with fewer now saying God created human beings in their present form and more saying humans evolved without God's help from less advanced forms of life over millions of years. Combined with the one-third of Americans who believe God guided evolution, a majority of U.S. adults thus believe humans evolved, yet a different majority still believe God played at least somewhat of a role in humankind's existence. (12)

At no time has Bible-literalist creationism been a majority public opinion in the USA.

Almost daily, creationists in social media groups will insist that mainstream biologists are abandoning the TOE in favour of ID, which, any day now, real soon, will become the prevailing scientific consensus. Meanwhile there are still "...no peer reviewed articles by anyone advocating for intelligent design supported by pertinent experiments or calculations which provide detailed rigorous accounts of how intelligent design of any biological system occurred."

What's Wrong With Intelligent Design?

I don't intend to spend too long refuting intelligent design, as I did just that with my book, *The Unintelligent Designer: Refuting the Intelligent Design Hoax* (13).

It is essentially the usual creationist combination of the presuppositional apologetic argument from incredulity and the God of the Gaps, that passes for scientific debate in creationists circles. It depends on the scientific illiteracy of its target audience combined with their unwillingness or inability to fact check and understand the science.

It also relies heavily on the assumption that complexity is a hallmark of intelligent design because only someone or something highly intelligent could design a highly complex system. And yet, the hallmark of good intelligent design is minimal complexity; the hallmark of a mindless, utilitarian evolutionary process operating without a plan and no mechanism for going in reverse or starting again, is what we see – layers of complexity evolving to improve on a suboptimal system.

The ID movement traces its origins to Michael J Behe's book, *Darwin's Black Box: The Biochemical Challenge to Evolution* (11) which pandered to the American creationist obsession with disproving Charles Darwin's 'Origin of Species'.

In it, Behe introduced the concept of 'irreducible complexity' in which he argues that if a biological system only functions as a whole structure, it could not have evolved by Darwinian gradualism.

However, the example he chose – the flagellum of the bacterium *Escherichia coli* (*E. coli*) not only exists in various different forms, some less complex than that of *E.coli*, but can easily be explained as the evolutionary exaptation of a redundant structure – the Type III secretory system (T3SS), used by some species of bacteria to attach themselves to their victims by extruding a protein filament. Thus, all the components of the flagellar motor were present having evolved for a different purpose by Darwinian gradualism.

But there was another even more embarrassing element to Behe's irreducible complexity argument, using *E. coli* as his subject, which creationists have been jubilantly waving around as 'proof' that the locally-popular god exists, and that is that *E. coli* can be a dangerous pathogen, so what Behe was arguing was that creationism's creator god creates pathogens, which appear to have a single purpose – making us sick.

In other words, creationism's creator god is a malevolent designer – theme I expand on in my book, *The Malevolent Designer: Why Nature's God is Not Good* (14)

Behe repeated the same mistake when he published another example of what he claimed was irreducible complexity in his book, *The Edge of Evolution: The Search for the Limits of Darwinism* (15) (again that obsession with disproving or diminishing Charles Darwin).- antimalarial drug resistance in the organism that causes malaria in humans, *Plasmodium falciparum*. Another example of creationism's intelligent designer god designing something to make us sick!

Basically, the mechanism of chloroquine resistance involves five component proteins, all of which need to be present for resistance to work. Behe argued that the probability of all five proteins arising together was minuscule (the probability argument we'll meet later on).

But Behe was taken to task by Professor Kenneth Miller, who showed that Behe had misused statistics to arrive at a minuscule probability of the system arising by assuming it happened as a single event in a single cell -which is not how evolution works in a population gene pool. A subsequent paper in Proceeding of the National Academy of Science (PNAS) (16) also showed that the system could have arisen in the *P. falciparum* gene pool by a number of different routes, reducing the probability of it evolving to a perfectly respectable number in a population of tens of millions where the million to one chance happens frequently. Bizarrely, Behe then claimed this paper vindicated his claim, when it flatly contradicted it and showed he had used the wrong mathematical model to arrive at his vanishingly small probability.

But again, here we have the archpriest of ID, in effect arguing that creationism's designer god works to make us sick.

So, the ID movement is now saddled with 'evidence' that their designer god is a malevolent sadist actively working through the parasites it designs to make us sick and find ways around the treatments medical science produces by designing anti-biotic resistant pathogens. It also participates in arms races with our defences and designs strategies to neutralise or circumvent the antibodies our immune systems produce. And, if that silly idea was not daft enough, creationists also believe the same designer god created the very immune system it takes on in these arms races. In effect it has arms races with itself, treating the solutions it designed yesterday as problems to be solved today! Creationists regard this as evidence of a high intelligence at work!

So, stuck with the notion of their designer god being the malevolence behind parasites, Behe obligingly wrote yet another book, *Darwin Devolves: The New Science About DNA that Challenges Evolution,* (17) in which he abandoned the pretence that ID isn't a religious movement and resorted to Christian fundamentalism, Bible literalism and some embarrassingly bad science to try to explain it.

In this book, Behe introduced creationists to two new sciencey sounding terms that they now regurgitate on cue when presented with evidence of parasites, pathogens, and genetic disorders – 'Genetic entropy' and 'Devolution'.

Genetic Entropy.

Genetic entropy is based on the notion taken from the Second Law of Thermodynamics (2LOT) that, in a close system, entropy tends to increase. Entropy is the amount of disorder in a system.

However, this analogy falls down when the full 2LOT is properly understood, particularly what a 'closed system is. In a closed system there is no way energy can be exchanged with the surrounding environment. But Earth and biological systems are not closed systems. The entire point of eating and breathing is to take in energy which is used to resist entropy.

Mutations are not an increase in genetic entropy, they are simply a rearrangement of the order in DNA which is maintained by replication and error-correction mechanisms which are fueled by metabolic processes using the energy in food.

Behe then argues that parasites, pathogens and genetic disorders are the result of genetic entropy since the initial created perfection! And that this was only made possible when 'Sin' entered the world, following Adam & Eves Original Sin. So, inserting Christian fundamentalism into his notion.

Behe doesn't explain how a 'perfect' system could change by a process of imperfections being accumulated. How can a perfect system have variances when it is replicated?

So, the first term can be dismissed as scientific gobbledygook intelligently designed to look nice and sciencey and so impress people who don't understand either physics or biology and for whom even joined-up thinking is problematic.

Devolution

We then have the biologically nonsensical notion of 'devolution' in which these inherited and accumulated imperfections are passed on to offspring which become increasingly imperfect over time.

However, when we look at the parasites and pathogens that have allegedly devolved, we see accumulated improvements which have been inherited; we see perfectly standard evolution, in fact. A bacterial pathogen which can circumvent an immune response is an improvement on a predecessor which couldn't.

Unless we have a definition of 'perfection' as something which can be improved upon, there is no logical way a beneficial mutation can be regarded as less perfect than what went before it.

So, we can see the ID movement still pursuing its subversive 'Wedge strategy' while the fundamental problems and contradictions caused by its lack of a scientific basis are forcing it to look for even more bizarre solutions, and now needing to fall back on Bible literalist Judeo-Christian fundamentalism while still maintaining the fiction that it's real science.

In the coming chapters I'll look at some of the criticisms and misrepresentations of science that the creation industry uses in pursuit of the Discovery Institute's Wedge Strategy, to try to destroy people's confidence and trust in science; still trying to flog the horse that was sick when it left the stalls in 1990 and died before the first bend – creation 'science'.

How Do We Know How Old Earth Is?

Geologists and palaeontologists use a number of methods for estimating the age of rock strata and objects found within them. Creationists dispute any or all of them, usually from a position of complete ignorance of the methods used.

For example, creationists will frequently assume that all geochronology is done by carbon-14 dating, which they've been fooled into believe in either flawed or easily faked. In fact, fossils and rocks are never dated by carbon-14 dating which depends on the presence of carbon derived from the specimen in question and, because of the relatively short half-life of ^{14}C, is only accurate up to about 50,000 years.

Creationists tend to have a schizophrenic attitude to ^{14}C dating; When it appears to support them, it is definitive proof; when it doesn't it is flawed or can be faked so doesn't prove anything.

I'll have more to say on ^{14}C dating later but for now, I'll look at how rocks and geological strata are dated:

Some, but not all, geochronology techniques rely on the decay of radioactive isotopes (radiometric dating). A radioactive isotope is a form of an element which has more than the 'standard' number of neutrons in its nucleus. ^{14}C is carbon with an atomic weight of 14 whereas 'normal' carbon (^{12}C) has an atomic weight of 12.

For the uninitiated, the superscript numbers are the atomic weight of the isotope. An atomic nucleus contains protons which carry one positive electrical charge and neutrons which have no charge. Each particle weighs about the same, so the atomic weight is the total weight of all the particles in the nucleus. In the case of ^{12}C, this is 6 protons and 6 neutrons; ^{14}C has 6 protons and 8 neutrons. Each proton has a corresponding orbital electron in an electron cloud surrounding the nucleus which gives the element its chemical properties. Since the negative electrons always equal the positive protons, the atom itself is electrically neutral.

Different isotopes have the same chemical properties because they have the same number of protons and electrons, but their different atomic weights can give them different physical properties.

Radioactive isotopes are naturally unstable and prone to sudden decay to the 'normal' state or another radioactive form. Some isotopes, like ^{238}U (uranium-238) decay by losing protons from their nucleus as well as neutrons, by ejecting a helium nucleus (2 protons and 2 neutrons) (an alpha particle). This changes them into a different element altogether with different chemical properties, and sometimes there is a 'decay chain' which goes through a series of elements, ending in a stable isotope of a different element. For example, the U-238 decay chain, also known as the uranium series, is:

1. Uranium-238 (U-238) decays to
2. Thorium-234 (Th-234) (half-life: 24.1 days) decays to
3. Protactinium-234 (Pa-234) (half-life: 1.17 minutes) decays to
4. Uranium-234 (U-234) (half-life: 245,500 years) decays to
5. Thorium-230 (Th-230) (half-life: 75,380 years) decays to
6. Radium-226 (Ra-226) (half-life: 1,600 years) decays to
7. Radon-222 (Rn-222) (half-life: 3.823 days) decays to
8. Polonium-218 (Po-218) (half-life: 3.10 minutes) decays to
9. Lead-214 (Pb-214) (half-life: 26.8 minutes) decays to
10. Bismuth-214 (Bi-214) (half-life: 19.7 minutes) decays to
11. Polonium-214 (Po-214) (half-life: 164.3 microseconds) decays to

12. Lead-210 (Pb-210) (half-life: 22.3 years) decays to
13. Bismuth-210 (Bi-210) (half-life: 5.01 days) decays to
14. Polonium-210 (Po-210) (half-life: 138.4 days) decays to
15. Lead-206 (Pb-206), which is stable.

There is a similar decay chain for ^{235}U which ends in a different stable isotope of lead (^{207}Pb).

The half-life of some of those intermediate elements is so short that they are irrelevant in the measurement of the age of strata estimated to be several hundred million or billions of years old, so the important point from a geochronologists point of view is the end-point - stable isotopes of lead (Pb) which is a very different element both chemically and physically from uranium, and this forms the basis of one of the most accurate and reliable geochronology techniques- so-called U-Pb dating of zircons. The half-life of ^{238}U to ^{206}Pb is about 4.5 billion years (i.e. longer than Earth has existed) while that of ^{235}U to ^{207}Pb is about 704 million years, so a significant proportion of the original uranium isotopes will still be present even after billions of years.

Uranium-Lead (U-Pb) Dating.

Zircons are crystals of the element zircon, which form in the cooling lava of volcanos and so can be found in rocks which are the solidified volcanic lava and volcanic ash (volcanic tufa), which falls over a wide area, even being transported by winds in the stratosphere over the whole globe, so a major period of volcanism will leave a record everywhere, unless subsequently eroded.

Because of their different physical properties, uranium, including its isotopes ^{238}U and ^{235}U can be incorporated in the crystal lattice of zircons but not lead. So, all zircons start their life with a small quantity of uranium but no lead.

Now move on several hundred million years, and some of that uranium will have decayed to stable isotopes of lead which is now locked up in the zircon crystal lattice where it could only be because there was once a radioactive uranium atom in the lattice.

So, by picking out the tiny zircon by hand under a microscope, geochronologists can gather enough crystals to be analysed to determine how much of the uranium has decayed to lead, and this gives the age of the rock, lake-bed deposits, or sediment in which the zircons were found.

For the technically-minded, the simple formula for calculating the age of the zircons is:

$$t = \frac{1}{\lambda} \ln\left(1 + \frac{D}{P}\right)$$

Where:

t is the age of the sample.

λ is the decay constant of the isotope (related to its half-life).

D is the number of daughter atoms (e.g., lead).

P is the number of parent atoms (e.g., uranium).

ln is the natural logarithm.

For the $^{238}U - ^{207}Pb$ series this would be:

$$t = \frac{1}{\lambda_{238}} \ln\left(1 + \frac{Pb - 207}{U - 238}\right)$$

For more information on U-Pb dating, see my detailed article in *Creationism in Crisis – How We Know Earth Is 4.5 Billion Years Old.* (18)

Other than ^{14}C dating which I'll describe later, other radiometric dating methods are:

Potassium-Argon (K-Ar) Dating:

This is typically used to date the crystallization of volcanic rocks and minerals such as micas and feldspars. Potassium-40 (^{40}K) decays to argon-40 (^{40}Ar) with a half-life of about 1.3 billion years. K-Ar dating is effective for dating rocks and minerals with ages ranging from a few thousand to several billion years.

Rubidium-Strontium (Rb-Sr) Dating:

This is commonly used to date igneous and metamorphic rocks containing minerals like feldspar and biotite. Rubidium-87 (^{87}Rb) decays to strontium-87 (^{87}Sr) with a half-life of about 48.8 billion years. Rb-Sr dating is suitable for rocks with ages ranging from tens of millions to several billion years.

The fallacy of changing radioactive decay rates:

One commonly floated dismissal of these dating methods by creationists is to claim, with no evidence to support it, and with very little understanding of the implications of the claim for their other dogmas, that radioactive decay rates have changed over time, but the dating methods assume they have been constant over geological time.

But these same creationists will be trying to assert that Earth, together with life on it was created only 6-10,000 years ago! Which means by amazing coincidence, that every dating method used to calculate ages of tens, hundred, or thousands of millions of years, is out by just the order of magnitude to make 6-10,000 years just happen to look like tens, hundreds or thousands of millions of years according to the radiometric evidence being disputed.

The problem creationists have unwittingly manufactured for themselves here is the dogmatic belief that everything science reveals must somehow be force-fitted into the last 6-10,000 years, no matter the contortions of the mental gymnastics which have to be performed to maintain that belief.

But even if it were true that radioactive decay rates used to be much higher just 6-10,000 years ago, creationists who use that false argument have not understood what determines the half-life of essentially random and uncaused radioactive decay. Nor do they realise they are arguing against one of their own sacred tenets – that God created everything just a few thousand years ago.

Radioactive decay is due to a random quantum fluctuation in energy levels sufficient for the ejected particle to overcome the weak nuclear force that binds the nucleus of an atom together. For decay rates to have been higher in earlier times (and creationists insist the maximum permitted age for anything is 6-10,000 years) the weak nuclear force when the rock was formed would have to have been even weaker, lowering the threshold at which radioactive decay occurs.

But that would mean atoms could not have formed! So, what creationists are actually arguing is that when they believe God created everything, including rocks and living organisms, the very atoms of which they are made could not have existed!

The same creationists will also claim the Universe was created perfectly tuned for the existence of life – the so-called 'fine-tuned-Universe – pointing out that if the fundamental forces had been slightly different, life could not exist. I'll say more about the fallacies in that argument later, but here I'll simply point out that they can't have it both ways. If radioactive decay rates were higher in earlier times, when they believe a god created life, life does not require the Universe to be fine tunes since one of the fundamental forces is the weak nuclear force that determines the rate of radioactive decay.

This is a perfect example of how ignorance of basic science enables creationists to hold two mutually exclusive views simultaneously without realising it.

It's a measure of the effects of Morton's Demon that, no matter how often you point that out to a creationist, they will keep repeating the lie that radioactive decay rates have changed so all radiometric geochronology will give false results., and the 'fine tuning' of the Universe proves a god must have designed it.

But even that scientifically nonsensical dismissal can't be applied to other, non-radiometric dating method, some of which are more accurate than others and serve different purposes. Some of these are:

Luminescence Dating:

This measures the quantity of trapped electrons in minerals like quartz and feldspar, which accumulate over time when exposed to sunlight or heat. The common types of luminescence dating include optically stimulated luminescence (OSL) and thermoluminescence (TL).

The method depends on the fact that electrons in the crystal lattice of minerals like quartz and feldspar can be excited by electromagnetic radiation and remain in that excited state until exposed to light and/or heat.

Electromagnetic radiation comes from the general background radiation from radioactive decay of elements such as uranium and cosmic radiation from the sun.

So, after they were last exposed to heat or light, these excited electrons accumulate at a steady rate, until again exposed to heat (thermoluminescence dating) or light (OSL) when the excited electrons return to their ground state and emit their excess energy as a photon. The amount of light given of is a measure of for how long excited electrons have been accumulating.

Palaeontologists use luminescence dating to date the last time sediment grains were exposed to sunlight or heat, providing ages for sediment layers containing fossils. (Note how the fossil is not dated directly but is inferred from the age of the sediment it formed in. Thermoluminescence dating is also useful for dating artifacts such as ceramics to determine when they were fired.

In order to mitigate possible sources of error, collection methods include measures to reduce further exposure to sunlight and laboratory procedures to accurately measure the luminescence signals and corrections to reduce dose rate variances. Dose rate variance is due to variability of background radiation and the fact that the contribution of cosmic radiation reduces with the depth of the sample.

Analysis of multiple samples is used to provide a statistical basis for estimating experimental error to give an average age with a known variance and a degree of confidence.

Calculation of the age since last exposure to sunlight or heat is then the simple formula:

$$Age(years) = \frac{D_e}{D}$$

Where:

D_e is Equivalent Dose (also called the Paleodose), which represents the total radiation dose absorbed by the sample since its last exposure to light or heat, measured in grays (Gy).

D is the Dose Rate, which is the rate at which the sample has absorbed radiation over time, typically measured in grays per year (Gy/year).

Paleomagnetic Dating:

Earth's magnetic field is subject to regular fluctuations which is recorded in the alignment of magnetic particles in rocks as they solidify. Known periods of magnetic reversals, as recorded in, for example, the magma welling up in mid-ocean ridges due to seabed spreading due to plate tectonic movements. These can be correlated with the magnetic alignments found in sedimentary rocks in which fossils are embedded. Again, the fossils are not dated directly but by dating the sediment in which they formed.

The limitation of this is that errors can arise from disturbance in the original rocks, variations in local magnetic fields and in the precise timing of magnetic reversals. Researchers are therefore careful to select samples from undisturbed sites and in the preparation of detailed geomagnetic records. Statistical techniques can be used to calculate the degree of uncertainty in the method which can be expressed as a range of dates with an upper and lower limit of confidence, rather than an absolute date.

Biostratigraphy:

Biostratigraphy uses a range of 'index fossils' of known age to infer the date of the rocks in which they are found. This it traditionally dismissed by creationists who have been misled into assuming scientists are stupid or dishonest enough to assign arbitrary dates to the fossils then, by ludicrously transparent circular reasoning, use those arbitrary dates to give the dates they want.

This of course, as with so much creationist propaganda, is at variance with the truth. In actual fact the fossils are dated by other methods first and only those which reliably give consistent dates in different strata are used as index fossils. It also uses something that creationists hate to accept – that there are series of fossils, especially micro-fossils, which show graduated change over time, so the exact form of the fossil can be used as an accurate measure of the age of the rocks.

In order to mitigate sources of error, palaeontologists compare the fossils with an assemblage of sample fossils, to allow for possible misidentification and distortions of the fossils under pressure.

Carbon Dating.

Carbon dating is probably the most well-known radiometric dating method and the one which seems to worry creationists the most since they spend an inordinate amount of time misrepresenting it and trying to discredit it – unless they think it supports the notion of a young Earth.

Carbon-14 (^{14}C) dating is, when done correctly on carefully decontaminated specimens, an accurate method of dating an organic specimen within well-known limits of confidence and within a well-defined date range. The date obtained will always be expressed as x years BP $\pm$ y years.

Doing it correctly not only means careful decontamination but also understanding its limitations, the major one being the date range over which it is reliable, why this limit exists and, importantly, exactly what carbon you are measuring in the sample and how it got there.

Not all carbon is organic in origin and not all carbon in archaeological samples is derived from the original organic carbon.

For example, although carbonates may be present in the mineralised bones of which fossils are made, this does not mean it came from the original bone. After all, fossils will contain large amounts of silicates, but bone does not normally contain silicon. The minerals in fossils are replacements for the original material, not necessarily the products of it.

The dating technique is based on three facts:

1. The isotope of carbon, ^{14}C, is present in the atmosphere at a more or less constant rate (but see below), being produced by the action of cosmic rays on nitrogen in the atmosphere.

2. Normal carbon is ^{12}C but, apart from being slightly heavier, ^{14}C is identical chemically to ^{12}C.

3. ^{14}C 'decays' at a constant statistical rate so that over a given period a known proportion of the ^{14}C will have 'decayed' to ^{12}N.

How Do We Know How Old Earth Is?

By comparing the predicted amount of ^{14}C in tree rings (dendrochronology) it is possible to estimate how the production of ^{14}C in the upper atmosphere changes over time due to variations in solar radiation and changes in Earth's magnetic field, so results can be corrected with these known variations.

When ^{14}C becomes incorporated into an organic molecule it becomes fixed in that molecule and so, unlike in the general background in which the amount of ^{14}C remains constant, the amount of ^{14}C in a sample of organic matter will fall over time. The known 'half-life' of ^{14}C is 5730 years so that, in 5730 years, half the ^{14}C will have decayed to ^{12}C.

So, by measuring the amount of ^{14}C in a sample and comparing it to the amount of ^{12}C, it is simply a matter of mathematics to calculate for how long it has been decaying, and so the age at which the original organic molecule was manufactured. However, the proportion of ^{14}C to ^{12}C in the general background is very small to begin with, being only about one atom in one trillion.

This means that a reasonably large sample is needed to begin with and, more importantly, it doesn't take long, no matter how large the sample, for the number of ^{14}C atoms to become too small to be significant. As the number of ^{14}C atoms present falls so the confidence in the measurement decreases until the range becomes so large as to make the calculated age almost meaningless.

For this reason, ^{14}C dating is only useful up to about 50,000 years and decreasingly so as that age approaches.

It also means that any contamination, especially from recent sources, can have a profound effect on the result making any sample appear to be much younger than it is. For this reason, ^{14}C dating is normally repeated several times and preferably by two or more independent laboratories. It is also verified by other dating techniques and 'corrected' for known changes in ^{14}C production.

It also means that care must be taken that the carbon being measured is actually the carbon that was present when the organic matter was made. For this reason, it is not used to date mineralised fossils. The process of fossilisation replaces the original organic matter with minerals, some of which may contain carbon, but carbon derived not from the fossil but from the environment, and this process can take place over a very long period including both before and after the specimen was living, so any inclusion of mineral carbon will distort the result by an unknow amount.

Creationists are understandably obsessed with ^{14}C dating because it shows how so much of the evidence that palaeontologists an archaeologists discover is much older that the 6-10,000 years allowed under creationist dogma. They are also acutely embarrassed by dinosaurs because they are not mentioned in the Bible and could not possibly all fit on a wooden boat for a year, and yet children are fascinated by them at just the age at which creationists need to be drawing them into the cult.

So, it's in the interests of creationist cult leaders to mislead people about both ^{14}C dating and how old dinosaurs are. Imagine what a propaganda coup finding a dinosaur fossil with enough ^{14}C in it to date it accurately to just a few thousand years old.

Well, surprise, surprise! That's exactly what a leading creationist apologist, Hugh Miller, claims to have done. In fact, he has a list of 'dinosaur fossils' which have returned ^{14}C dates of a few thousand years which is uncritically accepted by the creationist industry, but which have never been published in a peer-reviewed journal nor subjected to independent scrutiny and analysis.

Some of the fossils are in fact not even dinosaurs, Sample UGAMS-1935 is a bison, and the 'allosaur' (UGAMS-2947) is a mammoth. (19) The rest were 'identified' as dinosaur fossils by creationists.

Hugh Miller is a leading member of the Creation Research, Science Education Foundation (CRSEF), and the way he obtained these results is instructive.

What was ^{14}C dated was not carbon obtained from the 'dinosaur', such as collagen, but the fossil itself including any apatite. Apatite is a mineral derived from the bones and teeth of animals during the fossilisation process. It is a form of calcium phosphate:

$$(Ca_{5}(PO_4)_3.(F,Cl,OH))$$

It contains no carbon, but it can easily be contaminated by mineral carbonates from surrounding rocks, washed into the fossil by surface water. This is why a sample must be carefully decontaminated before being ^{14}C dated.

The whole point of ^{14}C dating is to date the carbon that was in the sample when it was living tissue in dynamic balance with atmospheric CO_2 and carbohydrates in their diet, as the source of ^{14}C. Measuring the ^{14}C in the contamination is bound to give erroneous results which is why the technique is never used by serious palaeontologists to date fossils, which are highly unlikely to contain much in the way of carbon derived from the living organism anyway, as any expert reviewer would have pointed out at the peer-review stage.

In fact, it looks very much as though the method used was designed to make the fossils look very much younger than they were.

But Miller had the results he needed to convince creationists who, while complaining that ^{14}C dating is flawed, will wave these results around as 'proof' that dinosaurs recently coexisted with humans on a 6-10,000-year-old planet, not realising that one of their objections to ^{14}C dating – the risk of contamination – was exploited to the full to obtain these results.

So, having dispelled the common creationist myths about their perceived fallacy of all the geochronology technique used by geologists and palaeontologists to arrive at dates for rocks and specimens that show earth and life on it predate creationism's supposed 'Creation Week' by several orders of magnitude we can move on now to their other obsession – with proving today's biodiversity was not the result of Darwinian evolution.

Underlying their constant attacks on the Theory of Evolution is the curious, but typical of creationists, assumption that by proving science is wrong, even in one tiny aspect of it, the whole of science is destroyed, and their superstition wins by default, without ever having produced a single supporting scientific datum.

In the next chapter, I will look at some of the common complaints by creationists about what they imagine evolution is and why they think it doesn't happen.

Evolution Works and Why Creationists Think It Doesn't

Any brief reading of creationist objections to the Theory of Evolution (TOE) will quickly show they think evolutionary biologists believe evolution is something that no sane biologist would ever think it is – the sudden transformation of one species, or an individual of that species, into a completely different taxon, so it will be essential for this chapter to have a clear idea of exactly what biologists mean by evolution and the different mechanisms that biologist accept as the underlying processes that lead to it.

Firstly, evolution is:

> Change in allele frequency in a population over time.

Nothing more and nothing less.

Because of occasional copying errors, genes can vary in their DNA sequences. The different versions of genes that this produce are known as alleles.

Allele frequency refers to the proportion of a population carrying that alle. So, any change in that frequency over any time scale is evolution in that population.

It also places the gene at the centre of evolution in the total gene pool of the population in question.

It is a principle of taxonomy that every descendant taxon is also a member of the ancestral taxon, so the evolved descendants of, say, the stem vertebrates, will always be vertebrates and the descendant of the last common ancestor of humans and the other apes will always be apes.

So, when creationists, as they frequently do, demand evidence of one 'kind' ever giving birth to an entirely different 'kind' – for example a cat giving birth to a dog – they are demanding evidence, not of evolution nor anything the TOE predicts, but of a childish parody, which, if ever evidence was produced for it, would falsify the TOE.

Creationists are fed this childish parody of evolution because the cult leaders know there can never be any evidence for it because it is not something that ever happened.

Creationists have also been fooled into thinking an entire species has to evolve into a different one for the TOE to be true. This works as a strategy because they are ignorant of the details of how evolution works, and how it is a population gene pool, not the entire species, nor individuals within it, that evolves.

Ignorance of the theory they are attacking is an essential part of creationism, maintained by Morton's Demon whom we met earlier, sitting at their door of perception and preventing any information which might shake their belief from reaching their consciousness, but allowing through the disinformation and childish parodies that reinforce their prejudices.

So, what are the mechanisms by which evolution occurs and what are the scientific principles that creationists would need to prove to be impossible in order to falsify the TOE.

There are four important mechanisms by which evolution occurs, so to refute evolution, all creationist need do is refute them all by explaining which laws of physics and/or chemistry prevent them happening.

The mechanisms, which I will consider in turn are:

1. Darwinian Natural Selection.
2. Genetic Drift.
3. Hybridization.
4. Horizontal Gene Transfer.

Darwinian Natural Selection

Much of what Charles Darwin wrote was written in ignorance of genes or DNA, so what he and, Wallace (20), his co-discoverer of the TOE, proposed was that natural selection by the environment was the cause of evolutionary diversification. They knew nothing of genes so did not consider genetic drift or horizontal gene transfers and were working to a different definition of evolution to the one science uses today.

They were also writing in the mid-19th century when some words had a different meaning and a different connotation to the way they are used today, when the term 'race' can be offensive. What they were referring to by 'race' was what today we would call variety.

Creationists will often latch onto this subtle difference in language to infer that Darwin was a racist, and that somehow the Theory of Evolution is the basis of racism. The irony is that so-called 'Social Darwinism', i.e., racism given a spurious scientific gloss, is the preserve of the far right in politics – the very people who tend to support creationism and oppose evolution in particular and science in general. Curiously, Darwin's supposed racism and the 'Social Darwinism' it gave rise to, is the only aspect of his Theory of Evolution that the far right embraced with enthusiasm.

Darwin and Wallace realised that the environment could 'favour' certain 'races' and so they would come to predominate in the species population. Darwin expressed this in the title of his ground-breaking book, *On the Origin of Species by Means of Natural Selection, or the Preservation of Favoured Races in the Struggle for Life* (21). The underlying assumption was that there must be some mechanism for passing information from one generation to the next, but what neither Darwin nor Wallace knew about was DNA and the genes it contains, arranged in Chromosomes.

What this means, and what creationists seem unable or unwilling to accept, is that a trait (i.e. the expression of an allele or group of alleles) which is more successful in a given environment will tend to have more descendants, carrying those alleles.

In effect then, selectors in the environment are acting as a sieve to allow through more of some alleles and fewer of others, so, over time, each generation tends to accumulate more of the 'favoured' alleles and fewer of the less favoured alleles. Meanwhile, the environment might well change, and a new set of selectors will come into play, with the new frequency of alleles to work on.

Note, there is no mechanism for reversing this process or abandoning this pathway and starting again, because, by definition, any reversion to earlier alleles will be less 'favoured' than those the selection sieve let through.

So, creationists need to explain what exactly makes any of that impossible. They can't of course, because it is logical, observable and irrefutable. So, what creationists will routinely do at that point in any debate is one or more of the following:

- Go completely silent and break off the debate.
- Switch to an ad hominem attack.
- Change the subject altogether,
- Demand evidence for one of their childish parodies of evolution.
- Cite a Bible or Qur'anic verse.
- Try to hold science to an impossible standard by demanding to see a complete set of fossils showing transition over time, regardless of the fact that fossilisation is a rare event in most environments and most fossils are probably still buried deep underground.

An interesting example of Darwinian evolution by natural selection was published recently in *Proceeding of the National Academy of Science (PNAS)* (22) in which a team based in the

Autonomous University of Barcelona showed how a species of plant, *Brassica fruticulose,* has evolved two different ways to cope with the same environmental problem – high soil salinity.

Brassica fruticulose

Brassica fruticulose is a costal species common around the Mediterranean and is closely related to the cultivated cabbage family (*Brassica oleracea*), and rape (*Brassica napus*) and to mustard (*Sinapis alba*)

There is a clear advantage to it to be able to grow in soil which has recently been inundated by seawater, or which frequently gets drenched in sea spray, and contaminated by sodium salts. But what the researchers found was that two different populations of *B. fruticulose* near Barcelona have evolved saline tolerance in two different ways, involving different sets of genes.

In the northern population, from the Cap de Creus region, the plant's roots have a mechanism for blocking the transport of saline into the rest of the plant, but in the central population, salt tolerance is achieved by having mechanism in their leaves where the sodium is accumulated. Osmotic adjustment and compartmentalisation then allow them to tolerate high concentrations of this compound.

It is absurd to think an intelligent designer would use two entirely different solutions to the same problem in the same plant, but there is no reason why a mindless process of natural selection could not produce two different solutions in two different gene pools.

What this example illustrates is another component in the process of sympatric speciation (i.e. where a species diverges into two different species while inhabiting more or less the same environment. In actual fact, this could be regarded as an example of allopatric speciation (i.e. where a population of a species become genetically isolated from its parent population and then diverges from it). In the case of this plant, the two populations are separated geographically but still could, in areas where their distribution overlaps, interbreed.

So, let's consider what might happen if they did interbreed:

We could end up with plants which have neither of the adaptations for high salinity, in which case they would fail to survive in that soil; we could have plants with both mechanisms, in which case one of the mechanisms would be redundant and a drain on resources, so would gradually be removed by natural selection and the population would end up with the 'winner'. Or, we could have a situation where neither plant inherited a full set of genes so neither solution worked properly.

The only one of those possible outcomes which is likely to be beneficial in the long term is the one where one of the solutions is eliminated; the other two would produce plants with reduced viability, so there would be selection pressure to prevent interbreeding. These mechanisms are known as barriers to hybridization which arise as species are beginning to diverge and so accelerate the rate of divergence.

How these barriers to hybridization arise is an important aspect of speciation about which I'll have more to say later. They refute the creationist claim that interbreeding will prevent speciation by dilution of any advantageous genes.

Genetic Drift

Genetic drift is due to the essentially random nature of which genes in the genome get passed into the next generation where there are no environmental selectors and the genes are neutral (i.e., they have no function, or they and their alleles are equally beneficial)

Their frequency in the gene pool is then free to rise and fall according to random chance. This is technically evolution because it meets the definition of change in allele frequency over time, but it makes little or no different to the species.

If they are around long enough, some of these genes will eventually disappear as their frequency reaches zero or they may become fixed in the genome if their frequency reaches 100%.

It is possible to calculate how long on average one of a pair of neutral alleles is likely to either disappear or become fixed, depending on the initial frequency and the population size:

$$T_{fix} = 4N_e$$

Where:

T_{fix} is the average time (in generations) to reach fixation.

N_e is the effective population size.

For an allele to become extinct, the formula is:

$$T_{ext} = -\frac{2N_e}{p}\ln(p)$$

Where:

T_{ext} is the average number of generations to reach extinction.

N_e is the effective population size.

p is the initial frequency in the population (between 0 and 1).

Where the allele is completely neutral, the probability of it becoming fixed in the population is simply the initial frequency, so an allele which has a 10% frequency has a 10% chance of becoming fixed in the population gene pool.

And lastly where the allele is not strictly neutral the formula is:

$$T_{fix} \approx \frac{2}{s}\ln(2N_e)$$

Where:

> T_{fix} is the average number of generations to reach fixation.
>
> N_e is the effective population size.
>
> s is the selection coefficient representing the advantage (or disadvantage the allele provides,

in the latter case, if the advantage is strong, so s is large, fixation can be achieved much more quickly than if there is only weak selection or the allele is neutral.

(Thanks to AI ChatGPT4o for those formulae)

Note that in those formulae, the smaller the effective population size (N_e) the more quickly fixation or extinction will be achieved, so the effect of genetic drift will be much greater with a small founder population. From then on, the local population will have a different set of alleles for the local environment to work on and so it will tend to diverge genetically from the parent population.

Generation time also has a major effect on the timescale over which genetic drift operates.

For example, with a human generation time of (say) 25 years, and T_{fix} of 200 generations, it would take on average 50,000 years for a neutral allele to become fixed (i.e. 100%) in the population, whereas, with a generation time of 3 months (say for a rat) the same T_{fix} would result in fixation on average in 50 years. Imagine the fixation time for some single-celled organisms with a generation time in hours or minutes!

From those formulae it can be seen that genetic drift has a much bigger effect on evolution in a small population, such as a small founder population that became isolated, or a population that has gone through a narrow genetic bottleneck by being reduced to a small population.

In these populations, a loss of genetic diversity is to be expected due to some alleles drifting to extinction, even if there is no selection pressure to cause it. Genetic drift can even be stronger than selection of beneficial alleles if selection is weak, so beneficial alleles can disappear from the population genome.

Some examples are:

The Amish and Ellis van Creveld syndrome.

The Amish population of North America, where a small founder population has resulted in a higher frequency of the genetic disorder, Ellis-van Creveld syndrome, which causes dwarfism, polydactyly and other defects, than in the general population. There is no advantage to the carriers of this disorder, but genetic drift in the founder population caused it to increase. (23)

Northern elephant seal

The population of Northern elephant seals was driven to near extinction by over hunting. Although numbers have recovered there is now a loss of genetic diversity which could be a long-term problem for the species.

Cheetahs

A similar effect is seen in the African cheetah, which went through a narrow genetic bottleneck, being reduced to a small number of individuals at some point in its history.

This was probably about 100,000 years ago when cheetahs expanded their range, into Asia, Europe and Africa and Africa could have been populated by a small founder population (24).

It now has low genetic diversity, is susceptible to disease and is less resilient to environmental changes.

Laboratory fruit flies

There is experimental evidence for this with laboratory fruit flies *Drosophila* where isolated populations have been shown to diverge genetically in the absence of selection pressures. (25)

A combination of genetic drift and natural selection could explain the distinctive features and small stature of the extinct hominin, *Homo floresensis* whose remains have been found on the Indonesian Flores Island, where a small founder population of *H. erectus* could have resulted in a population of small people, especially if small size meant lower demands on limited resource.

Of course, we don't (yet) have their genome to examine so we can't know for sure what their genetic diversity was, but the probability is that they are descended from a small founder population which then became genetically isolated.

Hybridization.

Hybridization is when two (normally) related species interbreed. Many, but not all, hybridizations result in a sterile offspring which is a genetic dead-end and can play no part in the evolution of either parent populations. However, some hybridizations produce viable offspring which can result in an ingression of the genes from one species into the genome of the other, sometime with spectacular results so far as evolution and speciation are concerned.

There are several examples of hybridization producing a new species which I will deal with in the next chapter, meanwhile here is a small selection of instances where hybridization has served to insert genes from one species into the genome of another.

Homo sapiens and Neanderthals

In 2010, Nobel Laureate Svante Pääbo and a team of evolutionary anthropologists published evidence of an ingression of Neanderthal, *Homo neanderthalensis* , genes into modern *Homo sapiens* in recent evolutionary history (26). Since then, further work has shown that modern humans, especially those of Southeast Asis, Austronesia and Oceania also have genes derived from a mysterious third hominin, the Denisovans, known only from a finger bone found in a cave in Siberia and teeth found in Nepal, but with, as yet, not enough skeletal evidence to base a scientific name on.

Neanderthals and Denisovans are also known to have interbreed since Denisovan genes have been identified in Neanderthal genomes

It is believed that Neanderthal genes contributed to the ability of *H. sapiens* to move into the colder climates of Europe and Northern Asia and may have conveyed some immunity to endemic pathogens.

This is just one of many examples of related species interbreeding, especially as they were diverging and evolving humans in Africa formed diverging isolated populations for prolonged periods of time, only to remix and merge into a single population again later.

There is also evidence of a prolonged period of occasional interbreeding between early hominins and chimpanzees at the two species diverged. (27)

Although these examples of change in allele frequency over time, are interesting, and often have creationists frothing at the mouths, there are many more even more spectacular results of hybridization producing entirely new species, in that rare thing in nature – a speciation event. Evolution is normally a slow process over time and rarely the parody events of which creationists continually demand evidence.

They tend to be more common in the plant kingdom where barriers to hybridization are not so solid as they tend to be in the animal kingdom. Plants are also prone to polyploidy (i.e., where the offspring has double the genome of the parents, due to mistakes in the production of the gametes.)

Polyploidy As a Cause of Speciation

Polyploidy arises in plants when there is an increase in the number of chromosome sets, either through errors in cell division (mitosis or meiosis) or through hybridization between different species. There are two main types of polyploidy:

> Autopolyploidy: This occurs when an organism has more than two sets of chromosomes that all come from the same species. It often results from the failure of chromosome separation (nondisjunction) during cell division, leading to diploid gametes that fuse to produce polyploid offspring. For example, if a diploid plant (2n) produces diploid gametes (2n), and these gametes fuse, the resulting plant will be tetraploid (4n).

> Allopolyploidy: This occurs when chromosome sets come from two different species, typically through hybridization. In such cases, a hybrid between two species may have unbalanced or non-matching sets of chromosomes, but if chromosome duplication occurs, this can lead to a new polyploid species with a stable genome. Allopolyploidy is common in many plants and is a significant mechanism of speciation.

Polyploidy can result in a new species when the polyploid individuals are reproductively isolated from the parent species. This isolation can be produced by reproductive isolation or by ecological divergence.

If the polyploid individual cannot successfully interbreed with its parent species because of mismatched chromosome numbers leading to infertile offspring, this can create a reproductive barrier, effectively isolating the polyploid population which can then follow and independent evolutionary trajectory.

Polyploidy may also convey some ecological advantage on the individuals, in which case they may be able to exploit a different ecological niche, resulting in genetic isolation.

Polyploidy is particularly important in plants because it can lead to instantaneous speciation. Many flowering plants (angiosperms), including crops like wheat (a hexaploid), cotton, and potatoes, have evolved through polyploidy.

Polyploidy is not confined to plants and, although it is rare in mammals, sometime occurs in vertebrates such as amphibians and lizards.

Edible Frogs

Although not a different species, the edible frog, *Pelophylax kl. esculentus* is a fertile result of hybridization between the Pool Frog (*Pelophylax lessonae*) and the Marsh Frog (*Pelophylax ridibundus*), hence the addition of the "*kl.*" (for klepton) in the species name, to indicate that it needs to steal' genetic material from another species in order to breed.

It is believed that the Pool frog and the Marsh frog diverged due to population isolation during the ice age, but speciation didn't progress far enough to prevent hybridization. However, the hybrids, although fertile often produce malformed offspring so they have never emerged as a hybrid species. The population is maintained predominantly by female edible frogs mating, usually, with male Pool frogs and rarely with male Marsh frogs.

This situation has arisen because when they produce reproductive gametes, the chromosomes of neither the Marsh frog or the Pool frog 'cross over' and exchange genes. Instead, each chromosome retains its full set of genes. This means that the hybrid Edible frog (*P. kl. esculentus*) has one set of *P. ridibundus* chromosomes and one set of *P. lessonae* chromosomes, so, when they mate with either of the other two species, they can produce either a *P. kl. esculentus* or one of the other two species depending on which they mated with. There is also a trisomy form of *P. kl. esculentus* which has two sets of one parent species and one set of the other. These are technically able to breed with one another, but the populations tend to be short lived so cannot be regarded as a distinct species. They can mate with either of the parent species with the same results as interbreeding between diploid individuals except that the offspring may be diploid or triploid.

I am sure the creationist explanation for this in terms of intelligent design would be both simple and convincing!

There are a couple more examples of speciation by hybridization that are interesting and indisputable refutations of creationists claims that speciation can't produce new species which can be regarded as forms of macro-evolution, i.e. where the change in taxon includes new structures or processes.

The Marbled Crayfish

The marbled crayfish (*Procambarus virginalis*), evolved in a German aquarium and is the only known all-female, asexual species of decapod crustacean. It reproduces parthenogenetically by cloning itself, and is spreading at an alarming rate, outcompeting and exterminating other species of crayfish as it goes. It now threatens the existence of seven species of crayfish in Madagascar. The new species came about when two American slough crayfish (*Procambarus fallax*), imported into Germany for the aquarium trade, mated.

Since its discovery in Germany in 1995 the marbled crayfish has spread across Europe and into Africa. The European Union has now banned the species from being sold, kept, distributed, or released to the wild. It came to the interest of medical science because there were thought to be similarities in how it clones itself and how cancer cells clone themselves.

Although it contains about the same number of genes, the genome of the marbled crayfish is larger than the human genome, at about 3.5 million DNA bases.

Genomic analysis show that it is a triploid *Procambarus fallax* that has 276 chromosomes instead of the normal 184 of its parent species. It also has several alleles of the same gene instead of the normal one or two, which gives it greater ecological flexibility enabling it to spread rapidly in a new environment and it can eat just about anything organic from dead plant matter to insects and even fish. (28)

It reproduces by a unique (to crayfish) method which does not involve the waste of external fertilisation, but it produces eggs which are genetic clones of itself. If an entirely new reproductive strategy does not constitute an example of macro-evolution, then I am at a loss to know what does. Presumably, a creationist would immediately revise the meaning of the word and complain that the crayfish didn't become an elephant or sprout wings, or some other equally childish parody of evolution.

New Mexico Whip-Tailed Lizard.

The New Mexico whip-tailed lizard is a very similar example of speciation by hybridization producing a triploid, genetically isolated population that reproduces parthenogenically.

New Mexico whiptail lizard (*Cnemidophorus neomexicanus*) is one of about 50 reptiles that normally breed parthenogenically (i.e. without mating) to produce clones of themselves. All these lizards are female. In fact, although parthenogenesis is the normal way of producing new *C. neomexicanus* there is one less common way. It is the way the first ones were formed - by two other species mating to produce a hybrid. Genetics has shown that *C. neomexicanus* arose, and still arises occasionally, when the two bisexual species, the little striped whiptail (*C. inornatus*) and the western whiptail (*C. tigris*), which have overlapping ranges, mate naturally (29).

C. neomexicanus is the result of a triploid egg being produced by fusion of a diploid gamete (egg or sperm) from one species with a normal haploid gamete from the other species. The result is invariably a female. However, unlike the Edible frog, the triploid lizard does not need to mate with one of the parent species to produce more whip tailed lizards. In fact, like the Marbled crayfish, it does not need to mate at all, because it produces eggs which are clones of itself. As such a single female can quickly produce a colony of whip tailed lizards.

One interesting aspect of its behaviour however is that, although there is no intromission and no exchange of genetic material, the female lizards still go through a mating ritual including mounting and cloacal contact in what can only be described as lesbian sex.

The evidence is that the lizard which adopts the female role in the mating ritual tends to produce larger and more viable eggs, so there appears to be some benefit from the behaviour.

Again, much to the distress of creationists we have speciation by hybridization, an entirely new reproductive strategy and lesbian sex all in the same species.

It is enough to make a fundamentalist weep.

Horizontal Gene Transfer

Horizontal Gene Transfer is a process where genes from one species can be acquired by another species, so adding to their genome and giving them additional abilities.

New Genetic Information and the Second Law of Thermodynamics.

Horizontal gene transfer flatly contradicts creationist claims that the Second Law of Thermodynamics (2LOT) somehow forbids the creation of new genetic information, because it is a rough and ready mechanism by which organisms can do just that.

That creationist article of faith comes from a misrepresentation of Shannon Information Theory, devised by Claude Shannon, who compared information to energy. The 2LOT states that in a closed system the degree of disorder (entropy) in a system tends to increase and creationists insist this applies to genetic information, which is an ordered system, therefore new information would be an 'impossible' increase in order.

However, they ignore the important clause, 'in a closed system'. A living cell is not a closed system, nor is Earth. The only truly closed system is the Universe itself since it is wholly contained within itself. A glance up at the daytime sky on a cloudless day will reveal the source of energy on Earth as Earth basks in a constant stream of solar energy, coming ultimately from the nuclear fusion of hydrogen into helium with the release of massive amounts of energy in the form of photons and other elementary particles.

All it takes to drive the metabolic processes that construct a new sequence of DNA, in other words to bring a degree of order from the chemicals inside the cell, is a source of energy, and our cells, like those of every other organism, get their energy from the food we consume, or in the case of photosynthesising plants, from the glucose they make from carbon dioxide, water and sunlight.

It is rare in higher taxons, apart from instances of retrovirus incorporation into the genome, but common in single-celled organisms. Bacteria, for example, transfer genetic information in the form of plasmids, or small sections of DNA, even between unrelated species, so, for example, antibiotic resistance can be spread from one species to another.

In addition to plasmids, viruses known as bacteriophage (or simply phage) can incorporate some of the bacterial genome in their own genome, then transfer it back into a different species when it infects them.

Although, as I said, horizontal gene transfer is rare in higher taxons, it undoubtedly played a part in our remote, single-celled ancestry.

But there are some interesting examples of horizontal gene transfer in some higher taxons such as arthropods and plants.

A Fruit fly/bacteria hybrid

The fruit fly, *Drosophila ananassae* , is not just an ordinary fruit fly. In fact, it's not just a fruit fly. It is actually a hybrid but not even a bog-standard hybrid between related species. It is actually a hybrid between two entirely different organisms. It is a hybrid between a fruit fly and a bacterium.

This extreme example of horizontal gene transfer genomes is believed to have happened about 8,000 years ago.

Biologically, there is no real mystery here because horizontal gene transfer has been known to science for several decades and the bacterium involved, *Wolbachia* is a genus of common endoparasites or endosymbionts in many arthropods.

Some insects have even come to depend on it for some functions, such as reproduction in the tsetse fly, resistance to some viruses and avoiding being parasitised by parasitoid wasps. But what is unusual is for an entire genome, from what was presumably an endosymbiont, to be transferred to the host genome.

The entire *Wolbachia* genome is embedded within the fruit fly's chromosome 4, which suggests the mechanism was incorporation of the Wolbachia DNA rather than RNA in the manner of a retrovirus. (30)

For a simplistic black vs white thinking creationist, trying to fit this hybrid species into a biblical 'kind' must be a nightmare. No such problem exists for biologists who understand evolution and how genes can transfer between species.

Photosynthesising Green Sea Slugs.

The green sea slug, *Elysia chlorotica* has a symbiotic relationship with chloroplasts which it gets from the marine alga *Vaucheria litorea*. These chloroplasts pass though the slug's digestive system unscathed and are relocated into the slug's body cells by phagocytosis where they can live for several months.

However, the chloroplasts need to be supported the way they are in the algae in order to survive and it is believed the green sea slug has acquired the gene *psbO*, for making an essential manganese-stabilising enzyme by horizontal gene transfer from the *Vaucheria litorea* and incorporated it into its own genome (31).

Endosymbiosis

Complex (eukaryote) cells evolved by incorporating simple (prokaryote) cells such as bacteria to become cell organelles such as mitochondria and chloroplasts (in plants). Part of this process included transferring part of the bacterial genome into the eukaryote genome, making the shared metabolic processes more efficient and reducing the size of organelles own genome. Mitochondria now have a much smaller genome than their free-living rickettsia-like ancestors and chloroplasts have lost their original cyanobacterial genome altogether.

Parasitic plants

In 2019, a team led by Professor Claude W. dePamphilis of Huck Institutes of the Life Sciences, Pennsylvania State University, USA, identified "108 genes, plus 42 additional regions with host-derived transposon, pseudogene and non-coding sequences", that have been added to dodder's genome by horizontal gene transfer and now seem to be functional in the parasite, contributing to haustoria structure, defence responses, and amino acid metabolism. One stolen gene even produces small segments of RNA known as micro-RNAs that are sent back into the host plant, acting as weapons that may play a role in silencing host defence genes. The micro-RNA normally plays a role in as part of the epigenetic system of gene control in the host, switching genes off, when not needed. By coopting this micro-RNA, dodder is able to control its host's immune response.

18 of these functional horizontal gene transfers have also been acquired by plants of the parasitic Orobanchaceae family from the same host source. (32).

New Metabolic Pathways in Fungi.

The fungus, *Aspergillus nidulans*, which metabolises dead wood has acquired genes for producing an enzyme for digesting cellulose, from bacteria.

The fungus lives in an environment which is awash with bacterial DNA from dead bacteria and the plasmids they leave lying around, so this gene could have been taken up from that environment without needed a close relationship between the fungus and the bacterium. Once taken up, the new gene would have given the fungus a strong competitive advantage over others in the gene pool so would have quickly progressed to fixation in that population.

The Asian Longhorned Beetle

The Asian longhorned beetle, *Anoplophora glabripennis*, is an invasive species notorious for its ability to destroy timber. It has acquired the ability to digest lignin, the structural carbohydrate in wood, and also detoxify allelochemicals in plants by horizontal gene transfer from bacteria and fungi.

In 2016, a large international team or researchers which included Prof. Duane McKenna from the University of Memphis, and

Prof. Stephen Richards of the Baylor College of Medicine concluded that, "Amplification and functional divergence of genes associated with specialized feeding on plants, including genes originally obtained via horizontal gene transfer from fungi and bacteria, contributed to the addition, expansion, and enhancement of the metabolic repertoire of the Asian longhorned beetle, certain other phytophagous beetles, and to a lesser degree, other phytophagous insects. Our results thus begin to establish a genomic basis for the evolutionary success of beetles on plants.

Bdelloid Rotifers

Bdelloid rotifers as microscopic multicellular eukaryotes that only exist as females and so reproduce by cloning. They currently hold the record for being the beneficiaries of horizontal gene transfer from bacteria with some 10% of their genomes being acquired this way.

In a 2015 paper in BMC Biology, a team of researchers led by Timothy G. Barraclough of the Department of Life Sciences, Imperial College London investigated the genomes of rotifers to test the hypothesis that, if HGT occurs when broken DNA is repaired after a period of desiccation, the genomes of rotifers from habitats prone o desiccation should contain more HGT's than those from permanently wet locations. They analysed the HGT's from "four congeneric species of bdelloids from different habitats: two from permanent aquatic habitats and two from temporary aquatic habitats that desiccate regularly.

They concluded that, "Nearly half of foreign genes were acquired before the divergence of bdelloid families over 60 Mya [million years ago]. Nonetheless, HGT is ongoing in bdelloids and has contributed to putative functional differences among species. Variation among our study species is consistent with the hypothesis that desiccating habitats promote HGT.

Horizontal gene transfer could be the basis for evolutionary divergence in an order that reproduces parthenogenically, so doesn't have the genetic remixing and crossovers that are part of the evolutionary process in sexually reproducing species.

Those then are the four main mechanisms by which evolution can occur. I'll now look at hybridization in greater depth because this is an area where creationists seem to find most to object to, claiming it shows how evolution is only within a 'kind'.

Species, Hybrids and Kinds

The Biblical Kind and Why Biologists Don't Use it.

'Kind' is the nebulous description used in the Bible to distinguish between different animals.

It has proved to be so inadequate that science has had to devise the currently used taxonomic system based on the branching tree analogy. Nevertheless, creationists still struggle to force-fit biology into their system of 'kinds' and do so only by having a flexible definition which can change from a single species, through families and orders to entire kingdoms as the needs of the debate arise.

A glance at the Bible shows the difficulty creationists are labouring under:

> And God said, Let the earth bring forth grass, the herb yielding seed, and the fruit tree yielding fruit after his kind, whose seed is in itself, upon the earth: and it was so.
>
> And the earth brought forth grass, and herb yielding seed after his kind, and the tree yielding fruit, whose seed was in itself, after his kind: and God saw that it was good. (Genesis 1: 11-12)
>
> And God created great whales, and every living creature that moveth, which the waters brought forth abundantly, after their kind, and every winged fowl after his kind: and God saw that it was good. (Genesis 1: 21)
>
> And God said, Let the earth bring forth the living creature after his kind, cattle, and creeping thing, and beast of the earth after his kind: and it was so. And God made the beast of the earth after his kind, and cattle after

their kind, and every thing that creepeth upon the earth after his kind: and God saw that it was good. (Genesis 1: 24-25)

And these are they which ye shall have in abomination among the fowls; they shall not be eaten, they are an abomination: the eagle, and the ossifrage, and the ospray,. And the vulture, and the kite after his kind; (Leviticus 11: 13)

Every raven after his kind; And the owl, and the night hawk, and the cuckow, and the hawk after his kind, And the little owl, and the cormorant, and the great owl, And the swan, and the pelican, and the gier eagle, And the stork, the heron after her kind, and the lapwing, and the bat. (Leviticus 11: 15-19)

Even these of them ye may eat; the locust after his kind, and the bald locust after his kind, and the beetle after his kind, and the grasshopper after his kind. (Leviticus 11: 22)

I wonder if a creationist could peruse that list and tell me how many 'kinds' of bird there are when there are even different 'kinds' of owl. And where do penguins, ostriches, and the flightless New Zealand kiwi and kākāpō fit in the biblical 'kind' scheme?

The inadequacy of the Bible's 'taxonomy' can be gauged by trying to use it to classify the order of insects, lepidoptera which includes moths and butterflies. Are those distinct kinds or all lepidoptera 'kind'? The bible offers no guidance on the matter.

If they are two different 'kinds' are all butterflies the same 'kind' or can we subdivide the butterfly 'kind' into, say, Monarch 'kind', Red admiral 'kind', etc?

If the latter, what do we make of the evidence of speciation into the different European Vanessid butterflies (33), a sub-family of the *Nymphalidae*, such as the Peacock butterfly, the Red admiral, the Painted Lady, the Comma and the Large and Small tortoiseshell 'kind'?

How about the divergence between butterflies north and south of the Pyrenees into the Eurasian Speckled wood and the Iberian Speckled wood or the Eurasian Brimstone and the Iberian Cleopatra?

Such is the inadequacy of the Bible's primitive attempt to classify animals and plants, if only for the purposes of arbitrary food taboos, that it falls apart immediately we try to use it for any meaningful taxonomy.

Hybridization and Scientific Taxonomy.

One area where creationists will tie themselves in knots is over the issue of hybridization, because they have a childish parody view of evolution of one 'kind' giving rise to an entirely different 'kind'; a dog giving birth to a cat or a chimpanzee giving birth to a human. So what they are usually attacking is not evolution, but a straw man intelligently designed by their cult leaders to be so ridiculous that no sane person would believe it – therefore the scientists who do believe it must be mad!

But evolution is, with a few exceptions such as instances where hybridization gives rise to a new species altogether, a slow process of gradual divergence over time, driven by local environmental selectors. There is no plan to produce new species; that they arise at all is a matter of chance,

This means that there will be a period of time when successful interbreeding is still possible. Creationists will point to those and proclaim it as proof that there has been no evolution, just variations of a 'kind'. Horse and donkeys can interbreed, albeit with sterile offspring, because they are both 'horse kind'; lions and tigers can interbreed, again with sterile offspring, because they are both 'cat kind', and yet they never stop to consider why the offspring are sterile.

The offspring are sterile because the parents are from species which have diverged to the point where they are effectively different genetically isolated gene pools. Interbreeding no longer leads to an exchange of genetic information between the diverging species, so, in any sensible definition of species, they have reached the status of distinct species.

Other diverging species have not (so far) achieved that status because nothing has driven it to that degree of difference. In North America, wolves can and do interbreed with coyotes, for example and, where their ranges overlap, North American brown bears and polar bears will interbreed, the polar bear only recently in geological time, having diverged from Brown bears. Occupying different habitats, they were unlikely to interbreed until climate change allowed Brown bears to move further north.

Creationists will claim this proves Brown bears and Polar bears are both 'bear kinds' while ignoring the fact that another bear, the Panda, does not interbred with other bears. So, is it a 'panda kind' or a 'bear kind'?

Scientific taxonomy does not have any problem with that, nor the fact that some related species can interbreed while being classified as distinct species.

The following is based on an article I wrote in my blog in an attempt to explain this 'problem' using the Eurasian Carrion crow/Hooded crow complex:

Carrion Crow/Hooded Crow

As you drive north from England across southern Scotland and up towards the Highlands, you might, if you're interested in the birds, notice that the ubiquitous glossy black crows you will have seen almost everywhere from town parks to country fields and woodlands have quite suddenly been replaced by an equally common crow with a black head, tail and wings and a grey back and underparts.

Species, Hybrids and Kinds

If you drive eastwards from Western Europe through Germany towards Poland or down into Austria or the Balkans, you will see a similar change. In both cases the plain black carrion crow, *Corvus corone*, has been replaced by the black and grey hooded crow, *C. cornix*. A similar change occurs as you drive from France into Italy south of the Alps.

What you may not have noticed as you drove north in Britain however was a narrow band where the crows were both black and grey and black, and an entire range of intermediates between the two. In this narrow band, which has moved over time, the two species of crow behave like a single species and interbreed freely, producing all the different intermediates.

There are similar zones of interbreeding between France and Italy. Everywhere else in Europe, they behave like two perfectly respectable species, just like, say, the song thrush, *Turdus philomelos*, and the blackbird, *T. merula*, or the house sparrow, *Passer domesticus*, and the tree sparrow, *P. montanus*.

If you continue driving eastwards from Europe across the Yenisei River into central Asia, you will find another sudden change - the hooded crows will disappear, to be replaced by what looks like slightly larger carrion crows. In fact, most authorities think that's exactly what the eastern crow is, and call it *C. c. orientalis*, the carrion crow being called *C. c. corone* to show that they are merely subspecies and not distinct species in their own right. Others disagree and think that, because the two subspecies are geographically separated, and have been for a long time, they do not interbreed and so form a single species with a single gene-pool, they should be regarded as two distinct species, *C. corone* and *C. orientalis*. However, there is a zone of interbreeding between *C. cornix* and *C. C. orientalis* just as there are zones of interbreeding between *C. cornix* and *C. c. corone*.

And it gets worse! Go from the north down through Iraq towards the Arabian Gulf and, as you come to ancient Mesopotamia between the Tigris and Euphrates rivers, the grey parts of the hooded crows become much paler, so they look almost black and white. There they are known as the Iraqi pied crow. Taxonomists regard these as a subspecies, *C. c. capellanus*. All in all, there are four subspecies of hooded crow; the other two being one which lives in a band running from western Siberia down between the Black and the Caspian seas through the Caucasus Mountains and into Iran (*C. c. sharpii*), and one found in Turkey and Egypt (*C. c. pallescens*).

The current hypothesis is that several populations of the stem species for all these crows got separated during the Ice Age and for tens of thousands of years evolved in isolation as isolated gene-pools. As the ice retreated one of the populations which had evolved into the hooded form expanded northwards forming a block between the western and eastern forms which had retained their all-black plumage.

These eastern and western forms have since continued on their different evolutionary trajectories, the eastern form becoming a little larger and their tail feathers becoming more tapered. Meanwhile, the hooded form, which happens to occupy a more diverse habitat has formed local subspecies. All the crows are sedentary species, never straying far from the area they were born in. This helps to minimise any gene flow from adjacent subspecies.

So why is this? Why does the science of taxonomy find it difficult to tell if closely related species are the same species, subspecies, or distinct species? Why is there now agreement that hooded crows and carrion crows, which do interbreed in a few small areas, are distinct species, yet disagreement about the status of the eastern crow, even though they never normally meet carrion crows in the wild and so do not normally interbreed? And, if the hooded and carrion crows were merely subspecies, what would be the status of the four subspecies of hooded crow?

Species, Hybrids and Kinds

Bible 'taxonomy' would have no problem, although none of these crows gets a mention. The only 'crow' in the Bible is the noise a cock makes in the morning.

The blunt instrument creationists use would simply declare them all to be 'crow kind' and having 'solved the problem' would ignore the differences and what they can tell us of the history of these birds and of Eurasia. Simple! 6-10,000 years ago, they were magically made that way from dirt and have been like it ever since.

The reason for this apparent confusion and imprecision in scientific taxonomy, is that 'species' is a human concept; a tool used by science to classify all the different living things. Nature does not read our rulebook and does not have any obligation to produce neat divisions between living things. Nature is quite happy about the distinction being blurred so any device we produce to try to divide living things up into our neat compartments is bound to lack precision because nature itself lacks precision. The definitions are fuzzy because the reality is fuzzy.

The reason *C. corone* and *C. corvix* are now regarded as species rather than subspecies, incidentally, is because studies have shown that the hybrids lack breeding vigour, indicating that divergence of the species had progress towards the point where interbreeding would either be impossible or the offspring would be sterile. This is in fact a departure from the commonly accepted definition because, although they may lack vigour, they may well be fertile, and one definition of 'species' is a distinct population capable of interbreeding and producing fertile offspring.

It was Charles Darwin himself who pointed out that if it hadn't been for evolution, which has caused living things to diverge and differentiate and which has emphasised and worked on differences, converting minor differences over time into major ones, that we have all the different taxa from kingdoms to subspecies and varieties in the first place. Evolution has carved out the current taxa from what would otherwise be a confusing mass of undifferentiated life forms. In fact, it would not have been possible for living things to evolve any of their solutions to problems like where to get basic resources from even to make copies of themselves. Life could not have progressed beyond simple autocatalytic replicators without natural selection to work on the slight differences produced by imperfect replication.

What we have in the crow example above is an example of evolution in progress, just like a ring species. These different populations of crows are seen at the current stage in their evolution and diversification. They are obviously not the same species but their progress to separate species status has not yet progressed to completion. In some ways they act like one species; in other ways they act like subspecies and in still other ways they behave like distinct species altogether. There is no requirement on nature or the process of evolution to complete this process suddenly or even within a given time frame, or in front of an eyewitness. If it takes 10 million years, or ten thousand years then that is what it takes. If you expect to see it happen, you have a loooooong wait.

In the case of our own divergence from the common ancestor we share with the chimpanzees, recent evidence (27) has suggested that we may have interbred with chimpanzees for over a million years before finally diverging. The process of speciation started some 6.3 million years ago and took 1.2 million years to complete! Later in the evolution of *Homo sapiens*, we were still able to interbreed with our cousins, H. neanderthalensis and Denisovans, and maybe with *H. antecessor, H. habilis,* and *H. erectus* and other species yet to be discovered (or should they be subspecies?).

This exposes the lie behind the idiotic mantra which creationists chant when you show them some examples of recent evolution or indisputable evidence that evolution occurred in the very recent past, "That's micro-evolution. Macro-evolution is

Species, Hybrids and Kinds

impossible! No one has ever seen a new species arise!" No-one who knows anything about evolution would expect to see a species arise (except in the rare examples of new species arising by hybridization – which have been seen) because it takes tens or hundreds of thousand or even millions of years. Evolution is a process, not an event.

By what sane logic have creationist pseud-scientists concluded that evolution can proceed within a species and can happily account for the evolution of subspecies and yet conclude that progressing a little further to the status of species is impossible? The terms are purely human constructs and do not mandate or constrain the process of evolution in any way. Are we really expected to believe that when humans classify two crows as *C. c. corone* and *C. c. cornix*, evolution is perfectly adequate to explain how they differentiated from one another, yet when the same humans decide to re-classify them as *C. corone* and *C. cornix*, all that magically changes and the differentiation we previously accepted somehow becomes retrospectively impossible? Only someone who sincerely wants to be fooled could fall for that 'reasoning'.

The creationist frauds who have sold that lie to the gullible fools who eagerly buy their cures for cognitive dissonance really should be taken to task for releasing their unfortunate victims onto the Internet to make such fools of themselves with that simplistic nonsense.

And the crow example can only be understood as an evolutionary process. There can be no intelligent reason for creating different overlapping populations with different forms and behaviours and then allowing them to interbreed, simply to produce offspring with reduced vigour which soon die out when they interbreed with one another, in a zone which needs to be constantly replenished with new hybrids from the two parent species. Any creator which intended that outcome would be an idiot. Any creator which did it accidentally would be incompetent.

The same frauds are also selling this 'Intelligent Design' nonsense to the same gullible fools.

The example of these crows is an example of what can happen as species diverge but do not progress to complete speciation.

The examples of horse/donkey and lion/tiger interbreeding are examples of what can happen when the process of evolutionary divergence has included effective barrier to geneflow between what are now indisputably distinct species.

There are also examples of what can happen when hybridization produces offspring which can breed with one another but not with either of their parent species. In that case we have a new species arising as a single event, not a gradual progress. These tend to be found in the plant kingdom:

Before I outline a couple of examples, consider the following questions which I will return to later:

If you saw a new plant that was a hybrid between two related parents, would you recognise it as a new species? If not, would it make sense to keep asking for evidence of a new species being seen to arise? Is a species comprised of a single individual?

Red Shepherd's Purse

This example is found in the family of common garden weeds, which I have always known as 'Shepherd's purse', with the generic name *Capsella*. They have a characteristic heart-shaped seed capsule which gives them their English name. These plants are something of a garden pest because they can harbour the fungal root infection of the cabbage family known as club root disease.

Capsella grandiflora, like many species of plant, has a mechanism which prevents self-pollination or 'selfing', known as self-incompatibility. Loss of this mechanism is fairly common in some plants because if offers a short-term solution where pollinators are rare. In that situation the disadvantage of not exchanging genes with another individual can be outweighed by the advantage of being able to produce fertile seeds, so natural selection can favour it in the right circumstances and work against it in other circumstances.

A very closely related species, *Capsella rubella* (red shepherd's purse), however is self-compatible.

By comparing the genomes of the two species, a team of researchers working at the Max Planck Institute for Developmental Biology, Germany, (34) found that the two species diversified only 30-50,000 years ago and, because of the low number of different alleles in *C. rubella*, it must have gone through a very narrow genetic bottleneck and probably arose from a single incidence of selfing, probably in Greece. Its spread, and maybe its survival, was facilitated by the spread of agriculture in the area at about the same time.

So, human agency at the dawn of agriculture may well have created the right environmental conditions in which 'selfing' became an advantage. That allowed a new species to arise and prosper, and one which continues to be an agricultural weed.

Given the definition of a species as a genetically isolated population, how could a single plant from which the entire species has subsequently descended be regarded as a species? Of course, it can't be and this speciation event can only be inferred later when the initial event is long gone and may not be recoverable. In other words, we can only ever know there is a new species when there is a population of them.

So, when creationists ask why there is no record of speciation being witnessed, you know you are dealing with someone who is either genuinely ignorant of what evolution is and how a species is defined or is feigning ignorance as a deception to fool the audience, confident that they will probably be that ignorant too.

Welsh Ragwort

Welsh ragwort or Welsh groundsel, *Senecio cambrensis* is an allopolyploid hybrid between *S. vulgaris* or Common groundsel, and *S. squalidus* or Oxford ragwort.

Hybrids between those two species, *S. x. baxteri* are fairly common but normally sterile. However, sometime in the early 20th century an accidental doubling of the genome of *S. x baxteri* produced a fertile hybrid that can only breed with other allopolyploid *S. x baxteri* and a new species was born, now names *S cambrensis*.

It was discovered in 1948 by Horace E. Green growing at Ffrith, Flintshire, Wales and was first systematically described in 1955 by Effie M. Rosser of Manchester Museum. In 1982 *S. cambrensis* was discovered growing in several sites near Edinburgh, Scotland. This population is believed to have arisen independently of the North Walian population in about 1974 but had disappeared by 1993.

It has been reported, probably mistakenly, in Shropshire, England and more reliably at Wolverhampton, England. The Wolverhampton population now also seems to have disappeared.

It had been around earlier than 1948 as a specimen collected in Brynteg, Wales in 1925 was found in an herbarium.

How this came about is an interesting example of how human agency can accidentally produce an environment favourable for evolution, as in the case of the red shepherd's purse above.

The Oxford ragwort is actually a Sicilian species that grew on the volcanic slopes of Mount Etna. It was brought to Oxford as a specimen for the University's Botanic Garden - the first such garden in the world - before 1719 but escaped and became established in the city growing on walls where it became known as the Oxford ragwort. It was described by Carl Linnaeus in

1753, but it is disputed whether his specimen was from the Botanic Garden or from a college wall.

With the coming of the railways, with their cinder tracks and stone ballast, the conditions were created for wind-blown seeds of the plant to spread, aided in this by the draft caused by the moving trains, causing the wind-blown seeds to spread along the tracks where the environment was idea for them to grow. It quickly spread to all parts of the country where it could come onto close proximity with the 'native' species, *S. vulgaris.*

So, human agency unwittingly created an environment which was conducive to the creation of a new species of ragwort, having been dispersed from its native home to provide a specimen for an academic institution.

The next example is one that was recorded on the walls of their caves by Stone Age artists.

Evolution of a New Species Recorded in Stone Age Cave Art

This example is of the wisent, or European Eastern bison, which suddenly appears in the fossil record about 10,000 years ago. That fact will no doubt excite creationists until they discover the explanation – it evolved from a hybrid between two much more ancient species.

In 2016, geneticists working at the Australian Centre for Ancient DNA at the University of Adelaide, showed that even the Stone Age painters of the caves such as Lascaux Cave, France, captured speciation in progress. Their findings were published in *Nature Communications* (35) and two of the authors, Alan Cooper and Julien Soubrier, of Adelaide University, also wrote an article about it in the open access, online magazine, *The Conversation* (36).

What they had discovered was that the artists who painted the pictures of animals on the walls of caves at Lascaux (about

20,000 BP, and at Pergouset (<17,000 BP) recorded three distinct forms of cattle.

Incidentally, those cave paintings would be there if creationism's genocidal flood myth was true since they would have been destroyed by flood water.

Fortunately, there was no such flood, so we have a record of a speciation process in which the auroch, *Bos primogenus* – the long-horned ancestor of domestic cattle and the steppe bison, *Bison priscus*, hybridized to produce the fertile offspring, the Eastern bison or wisent, *Bison bonasus*.

What the cave painters had recorded was the auroch with its long horns and relatively light shoulders, the Steppe bison with its short, curved horns and heavy shoulders, and an intermediate bison, with the auroch's long horns and the heavy shoulders of the steppe bison.

This Eastern bison or wisent was to go on to become the dominant bison of Europe, interchanging that status with the Steppe bison as the climate fluctuated back and forth, favouring first one then the other. It was to produce a subspecies in the Caucasus, *B. b. caucasicus*, the Caucasian bison.

This hybridization event has left its mark in the DNA of the wisent, which shows close affinity with the American bison, but a closer affinity, particularly in its mitochondrial DNA, with modern cattle, indicating an ingression of ancestral cattle DNA at about the time the Stone Age artists were recording the three forms on cave walls.

The obvious question then is why, if hybridization occurs as species diverge, why do they ever diverge in the first place, especially where the populations are not physically separated?

The answer is that the same reasons that the genomes diverged in the first place become selection pressures to maintain that separation so there is pressure to evolve barriers to hybridization. This is a significant driver of speciation which I will deal with in greater length later. But first, the consequences for biodiversity if the pressure for these barriers to arise is weak.

Ring species

Of course, there are instances where this does tend to delay complete speciation, as in the Carrion crow/Hooded crow example above, and there are several examples of 'ring species' where speciation is gradual and progresses at different rates in different parts of it, to form a 'cline' which reflects a graduated environment across the range.

Larus gulls

A famous example is that of the Larus gull circumpolar complex.

Starting at the Eastern coast of the North Atlantic, around the British Isles and Scandinavia there are two related species of gull, the Herring gull, *Larus argentatus argentatus* and the Lesser black-backed gull, *Larus fuscus*. They behave like two perfectly normal species and don't normally interbreed.

However, as we progress eastward into Russia and across Siberia across the Bearing Straight into Alaska and Canada, we encounter some interesting changes in the local equivalent of these gulls.

Firstly, we find a sub-sub-species of Lesser black-backed gull, *Larus. f. sensu stricto*, which interbreeds with *L. a. argentatus*. This in turn is replaced in Northern Russia by another subspecies, *L. f. heuglini* , which can interbreed with *L. f. s. stricto*. Then we have in turn, each of which can interbreed with its neighbouring species, *L. a. birulai, L. a. vegae, .L. a. smithsonianus*, and finally *Larus a. argenteus*.

Clearly, what is happening here is that gene flow over such a long distance is not sufficient to prevent the gulls in the Eastern Atlantic from forming perfectly normal species, but nowhere in the circumpolar continuum have populations of local sub-species been isolated for long enough to form local species.

The Greenish Warbler

The greenish warbler (*Phylloscopus trochiloides*) forms a complex of sub-species around the Tibetan plateau. It is a member of a widespread group of 'leaf warblers', a group which is taxonomically close to the chats and thrushes which included both the European and American robins and nightingales.

Once again, we see a species diversifying either side of a geographical barrier, forming several sub-species which can interbreed with their neighbours, until they meet on the far side of the barrier, at which point, even though they co-exist, no interbreeding occurs, and they behave exactly like different species.

Ensatina salamanders of California.

This group of salamanders (*Ensatina escholtzii*) is believed to have evolved as the original species from Oregon and Washington extended its range southward to the San Joaquin Valley in California and into southern California. As the species moved down either side of the valley, separated by the valley floor, they occupied and adapted to the assorted opportunities and threats the new niches presented.

By the time they had reached southern California they had formed two quite distinct forms: one (*E. e. klauberi*) having dark blotches in a camouflage pattern; the other (*E. e. escholtzii*) less strongly marked and with yellow eyes which may mimic those of the poisonous western newt. These two sub-species coexist in some areas but do not, and apparently cannot, interbreed.

So, to a southern Californian taxonomist unaware of the situation further north, these two salamanders are distinct species, filling all the requirements of that classification. Yet, following each species north on either side of San Joaquin

Valley we find other varieties with which interbreeding occurs naturally until, at the northern end of the valley we find a single species.

What we have is speciation in progress with evidence of the intermediate forms preserved. All we would need to happen is that the subspecies either side of the valley went extinct and left no geological record and we would have speciation without known transitional forms - something which creation 'scientists' would then wave jubilantly as 'proof' of creation.

A Human ring species?

It can be argued that at least once in out evolutionary history, the *Homo* genus acted like a ring species complex. *H. erectus* migrated out of Africa first and spread across Eurasia, giving rise to Neanderthals in the Weast and Denisovans in the East, each of which could interbreed with the other. The descendants of the African *H. erectus* meanwhile evolved into *H. sapiens*, who then migrated up into Eurasia following the route previously used by H. erectus.

There they came into contact with their cousin species, Neanderthals and Denisovans and interbreed with them, receiving an ingression of their genes into the modern, non-African *H. sapiens* who then spread throughout the rest of the world carrying these archaic genes with them.

For a period of some 20-30,000 years there were three interbreeding *Homo* species in Eurasia behaving much like a ring specie and there is tentative evidence of a third, as yet unidentified hominin in that complex (37).

And in 2021, evidence of yet another hominin interbreeding with *H. sapiens* was found in the DNA of a female who was buried in the Leang Panninge Cave on the Indonesian island of Sulawesi, more than 7,000 years ago. (38)

There is also genomic evidence that the African hominins diversified to form local populations then merged again when they came back into contact. One group diversified from the main hominin population some 1.5 to 2 million years ago, to form a distinct population from which there was an ingression of DNA some 150,000 years ago.

The embarrassing thing for creationists is that this evidence of evolutionary diversification into different species which later interbred to produce the present population of *Homo sapiens* is so different to their myth of a founder couple with no ancestors who then committed the 'Original Sin' for which we all supposedly need to atone and seek forgiveness and redemption.

Not only was there no such founder couple, there was not even a founder species, so who and when committed this 'Original Sin' that has traditionally give the church so much power over us?

Barriers to hybridization

Barriers to hybridization come under two different headings, each operating at various levels in the reproductive cycle;

Pre-zygotic barriers

Pre-zygotic barriers operate before the formation of the zygote from a fertilised egg. This can take many forms depending on the species. For example, in some insects, mating is impossible because of the anatomy of the reproductive organs. In birds, especially, plumage, female sex selection and mating rituals all play a part in ensuring the female only mates with members of her own species.

In some plants, such as orchids, the relationship between the flower and the pollinator may ensure only pollen from the right species reaches the female part of the flower, or the wrong pollen grains may fail to germinate on the stigma.

Species, Hybrids and Kinds

Post-zygotic barriers

Post-zygotic barriers to hybridization are barriers that prevent a zygote producing a fertile offspring. These can take the form of genetic incompatibility so a zygote can't develop beyond the single-celled or blastocyst stage, or, as in the case of the horse/donkey and lion/tiger hybrids, the offspring is sterile and a genetic dead end that plays no further part on the history of the parent species.

In mammals, there is the opportunity for the female to reject the developing embryo with an immune response, while the colour displays of birds are not available due to most mammals being colour blind, unlike birds which have full colour vison, so in mammals, post-zygotic barriers tend to predominate whereas pre-zygotic barriers are more common in birds.

In addition, mammalian chromosomes are more liable to reorganisation in the form of breaking into two or fusing back into one, whereas bird genomes are very stable – normally between 28-30 chromosomes. so genetic barriers are more common in mammals than in birds.

Of course, these are not mutually exclusive and both pre-zygotic and post-zygotic barrier may play a part in speciation.

So, to get back to the original question, why, if diverging species can still share genetic material which prevents them becoming completely distinct species, are there so many species that did emerge by evolutionary divergence from a common ancestor.

The answer to that is to be found in the way barriers to hybridization form and why they form, bearing in mind that there is no underlying plan to produce new species.

The following is based on a blog post I wrote some years ago:

European finches

In the standard model, the first essential is that a group of individuals becomes isolated from the main population for long enough for gradual changes to accumulate in response to local environmental factors and genetic drift.

These factors may be predation, success at finding food, breeding success, etc. The main population will meanwhile be changing in its own way in response to its own local environment, or not, if the environment is stable.

Eventually difference may build up in each population so that, if ever they do come back into contact again the two populations' genetic make-up will be such that they physically cannot interbreed successfully to produce fertile offspring, even if they can still successfully mate. This is the case with donkeys and horses, lions and tigers and many species of plant.

In this model speciation is merely a passive, incidental result of gradual evolution. There is another model however, in which speciation is itself driven by evolutionary selections. (I will use European finches to illustrate this, but I could equally have chosen almost anything else, insects, reptiles, plants, fish or frogs, etc.)

Consider Europe either side of the last ice age. Northern Europe, the Alps to the north of Italy and the Pyrenees between Spain and France were all heavily glaciated, driving many species south into Spain, Italy and the Balkans and effectively isolating them there with impassable ice sheets.

Now take a species of finch, adapted to live in Northern Europe with a generalised bill for eating a variety of seeds. This would have been pushed south to form several isolated populations. Each would have evolved and adapted to best use the evolving and changing plant population.

Species, Hybrids and Kinds

One species in, say, Spain, may have evolved slender bills for picking seeds from thistles and other wildflowers, the more successful ones passing these bills on to their offspring. The other population in, say, Italy, may have evolved stouter bills for cracking harder seeds, also passing these on to their descendants. The two populations would be diversifying according to local selection pressures.

Now move on to the end of the ice age when the Alps and Pyrenees became free enough of ice for the finches to return, together with their food plants, into an increasingly temperate Europe as the ice retreated. Suppose these finches had not been isolated for long enough to make interbreeding impossible.

What type of bill would their offspring inherit?

They would probably inherit an intermediate bill, but an intermediate bill which was no use for either of the favoured food plants of its parents. To all intents and purposes, they would be handicapped and incapable of feeding or capable of feeding only with difficulty.

These would be rapidly removed from the gene-pool. Interbreeding would be hugely wasteful as the result of all that effort would be a lost brood, or, at best, a brood of individuals with a greatly reduced chance of themselves producing offspring.

Anything which favoured non-interbreeding between the two forms would now be highly favourable. Changes in plumage combined with display mating rituals, territorial and mating song, and, especially, female sex selection would all be favoured.

Gene-pool isolation would be reinforced now, not by geography but by any other means available. A process of speciation which began casually and incidentally in, and because of, isolation, would now be accelerated paradoxically by a lack of the very isolation which initiated it.

And so, we have lots of different related finches in Europe, each with its own plumage, song, mating rituals and food plants, many of which are actually STILL capable of interbreeding successfully, and do so in captivity, but which rarely do in the wild.

Speciation has occurred because it was in the 'interests' of both gene-pools to speciate. An incidental yet inevitable result of evolution and an undirected, yet highly directional, process of natural selection acting as though it were driven by the needs of genes to replicate through time.

The case of the European finches is an example of pre-zygotic barriers where plumage, song and mating rituals make it unlikely (but not impossible) for interbreeding to occur.

Pre-zygotic barriers can produce some interesting results especially where, once mating does occur, there are few if any post-zygotic barriers so a viable hybrid can result.

When pre-zygotic barrier fail – hybrid geese and ducks.

The geese and duck are a group of birds in which this happens frequently, to the extent that some native species of duck are being threatened by interbreeding with the introduced ruddy duck, *Oxyura jamaicensis*.

The following is based on an article I wrote in response to seeing a strange-looking goose on a local lake – clearly a hybrid between a native grey lag goose and the introduced Canada goose:

What we noticed about these geese was that, while they are obviously a pair and one was a perfectly normal graylag goose, *Anser anser*, the other was a slightly odd-looking Canada goose, *Branta canadensis*. Canada geese are an alien species in Britain but have spread very rapidly throughout the Thames Valley and beyond.

Species, Hybrids and Kinds

Graylags are, of course, a native species and the probable ancestor of the domestic goose. What was odd about the Canada goose is that instead of the normal black head and neck, with white cheek patches, this one was spotted with white where it should have been all black and its back looked a bit, well... graylagish (one doesn't often get an opportunity to use that adjective!)

The obvious conclusion is that the odd goose is a hybrid between a Canada goose and a graylag - hybridization if fairly common in the *Anatidae* group (swans, geese and ducks) especially where they occur together in quite large numbers, but why would a normal graylag pair up with a hybrid which looks far more like a Canada goose than a graylag? And why is hybridization so frequent in the *Anatidae*.

This is where it starts to get really interesting, to anyone who wants to understand the world around them.

There are believed to be two main causes of hybridization in this group, in which the males are unusual in birds in that they have retained a penis which nearly all other birds have lost in evolution (39).

The first is by, not to put too fine a point on it, rape. One reason males of this group of birds may have retained a penis is because male genes benefit from their carrier's ability to force a female to copulate. In species where sexual dimorphism is pronounced, as in most ducks, suggesting a high degree of female sex selection, there is an obvious survival advantage in the males having a strategy for overriding it. There is no morality in evolution.

The second is a sneaky strategy some females have, so they can produce more descendants in a year than they could raise themselves. They are partial brood parasites, like the cuckoo, and will lay eggs in another nest. Normally, this will be those of the same species, but not always, so their offspring can end up being reared by a different species.

Given that the young of this group are active from hatching and are not fed by their parents, it actually makes little difference to them who is keeping them warm at night and protecting them from predators.

But, geese, especially, 'know' what species they are by imprinting on the first moving thing they see when they hatch, so these 'cuckoo' goslings will think they are a different species and will seek out members of that species for a mate.

So, that is why we have frequent hybridizations in this group, and probably why our hybrid Canada/graylag goose was paired with a normal graylag, having been raised as a graylag, probably conceived by rape.

But now we have another question: why is stable, fertile hybridization so common in this group in the first place and why is it biologically possible when it's normally rare in mammals, and usually results in either non-viable or sterile offspring? Once speciation occurs there are distinct advantages to both sister species to evolve barriers to hybridization, as I illustrated with the European finch example.

Biologists classify these as pre-zygotic and post-zygotic barriers. Pre-zygotic barriers are ones which prevent egg and sperm ever coming into contact by preventing mating. In birds, the barriers often involve mating rituals and displays, involving colour. Birds have the advantage there over most mammals in having colour vision. Most mammals do not have colour vision, apart from the simian branch of which we and the other great apes are members.

Post-zygotic barriers are those which prevent the fertilised egg from developing. Normally, this will be because of genetic incompatibility in that the genes of one species are organised differently to those of potential mates, so, even if mating occurs and the egg is fertilised, a viable offspring will not develop. In mammals there can be an additional post-zygotic barrier in that the female's immune system might detect and destroy a 'foreign' embryo developing with a placenta.

This is not available to birds, however because the female's reproductive tract quickly seals the fertilized egg with a shell and expels it. For this reason, birds are much more dependent on pre-zygotic barriers and mammals on post-zygotic barriers. Hybridization is much more common in birds other than just the *Anatidae* than it is in mammals, and this partly explains the brilliant colours of many bird species.

So, that just leaved genetic incompatibility to explain away.

Typically, mammals will have a widely differing number of chromosomes between different species because the mammalian chromosomes seem to be more prone to fusing into one or breaking into two. Most mammal chromosome numbers fall within the range of 18-30 pairs (humans have 32 pairs). Most birds, by contrast, fall within a much narrower range of 38-40 pairs and all the geese have 40 pairs.

This evolutionary stability means there are few genetic barriers to hybridization either, even for species that have been geographically separated for tens of millions of years.

The result is that if a male can get through the pre-zygotic barriers he has a good chance of a successful, viable fertilisation and that his offspring will be fertile.

So, the combination of a retained penis, a rape strategy as a mode of mating, and the female's parasitic strategy for egg-laying, all conspire to help the male get past the pre-zygotic barriers and the lack of post zygotic barriers makes this strategy viable.

In evolutionary terms there have been two competing strategies – the pre-zygotic barrier involving plumage, female sex selection and mating rituals, and the brutal rape strategy of males, the former serving the female genes, the latter the male.

So, that is probably why we saw slightly odd-looking Canada goose paired with a perfectly normal graylag goose.

Somehow, I find this explanation hugely more satisfying than the simplistic and arrogantly dismissive "God did it! And He did it all for me!" answer you will get from any proudly ignorant creationist.

Macro-Evolution Vs Micro-Evolution

Definitions and Applications

First, brief definitions of the two terms, how they are used in biology and why the distinction is often blurred, provided by ChatGPT4o.

Yes, there is a scientific distinction between macro-evolution and micro-evolution, though the boundary between the two is more conceptual than sharply defined.

Micro-evolution

Micro-evolution refers to small-scale changes in a population's genetic makeup over a brief period of time. These changes typically occur at or below the species level and are driven by processes such as:

- Natural selection
- Mutation
- Gene flow (migration)
- Genetic drift
- Non-random mating (sexual selection)

These processes lead to changes in allele frequencies within a population. Examples include:

- Antibiotic resistance in bacteria
- Changes in coloration of peppered moths due to industrial pollution
- Variation in beak sizes of Darwin's finches in response to food availability

Micro-evolution is observable in real-time over generations and is often studied in populations of organisms with short life cycles.

Macro-evolution

Macro-evolution, on the other hand, refers to larger-scale evolutionary changes that occur over extended periods of geological time. These changes often result in the emergence of new species, genera, families, or even higher taxonomic levels. Macro-evolution involves processes that go beyond the scope of micro-evolutionary changes, such as:

- Speciation: The process by which new species arise, either through gradual divergence (as in allopatric speciation) or more sudden events (e.g., polyploidy in plants).
- Mass extinctions and subsequent adaptive radiations: Major ecological shifts, such as the diversification of mammals after the extinction of non-avian dinosaurs.
- Evolutionary trends: Long-term directional changes in traits across many generations, like the evolution of larger brains in some primates or the transition from land-dwelling mammals to whales.

Key Conceptual Differences

1. Scale of Change: Micro-evolution deals with small changes within species, while macro-evolution encompasses the broader patterns of evolution that lead to the formation of new species and higher taxonomic groups.
2. Timescale: Micro-evolution occurs over shorter timescales (e.g., a few generations), whereas macro-evolution unfolds over longer periods, often involving thousands to millions of years.
3. Outcome: Micro-evolution results in variations within a population, potentially leading to adaptation or shifts in traits. Macro-evolution describes the accumulation of these minor changes over long timescales, eventually leading to speciation and larger evolutionary transitions.

4. Processes: Both micro-evolution and macro-evolution are governed by the same basic evolutionary mechanisms, but macro-evolution includes additional phenomena, such as speciation and extinction, that shape the diversity of life at higher taxonomic levels.

Scientific Definition of Macro-evolution

While there is not a rigid or universally agreed-upon definition of macro-evolution, it is understood as:

> "Evolutionary change that occurs at or above the level of species, encompassing the origin of new species and the diversification of higher taxa, such as genera, families, and orders, over long geological time periods."

Evolutionary biologists agree that macro-evolution can be explained by the cumulative effects of micro-evolutionary processes. However, macro-evolution involves additional factors such as mass extinctions, large-scale environmental changes, and developmental biology changes (e.g., changes in regulatory genes that affect body plans).

The Controversy in Historical Context

In the early 20th century, some scientists viewed macro-evolution as distinct from micro-evolution, suggesting that different mechanisms were at work. However, today, the majority of biologists agree that there is no fundamental difference in the mechanisms driving both processes. Micro-evolutionary changes accumulate over time and, under certain circumstances, lead to macro-evolutionary patterns. The distinction is one of scale rather than underlying process.

Summary:

In short, micro-evolution refers to small-scale changes within a species or population, while macro-evolution describes large-scale evolutionary patterns that can lead to new species or higher taxonomic groups. Both are driven by the same fundamental evolutionary processes, but macro-evolution focuses on the broader, long-term outcomes of these processes.

Creationists embrace the concept of macro-evolution because they imagine it fits neatly into their biologically nonsensical notion of 'kinds.', so we can dispel that nonsense easily:

There are not two different sorts of evolution.

Challenge a creationist to explain the different processes involved in their two sorts of evolution and explain how they differ and you'll be met by the usual tactics of avoidance and diversion, but one thing you will never get is an answer to your question, because there is none.

Evolution above the level of the species taxon is by exactly the same underlying processes as at the species level.

What evolutionary biologists mean by 'macro-evolution' is a new taxon based on new morphological features, physiological processes, behaviour, and/or habitat.

Creationists need to deploy their flexible definition of 'kind' to dismiss the evidence of evolutionary divergence into different genera, for example. In the example of the hybrid goose, I used earlier, the two species involved in that hybridization were from two different genera – *Anser* and *Branta* and yet they were sufficiently close to be able to interbreed.

Taxonomists would regard them as examples of macro-evolution as they diverged from a common ancestor, while creationists would either argue that it must have been impossible because macro-evolution can't happen, or they will adjust the definition of micro-evolution so both genera are 'goose kind' and voilà! It is now micro-evolution, which of course is perfectly possible.

And yet, as we know, the scientific definition of evolution is a change in allele frequency in a population over time. There is not a different process that produces macro-evolution; all there is, is a series of micro-evolutions that eventually get classified in human taxonomy as a new taxon. Creationists would have us believe that there is some magical threshold at which lots of micro-evolutions come up against a barrier because one more would make it add up to a macro-evolution, so something in the laws of chemistry and/or physics bust be operating to enforce that barrier.

Like so much else in creationism, it is nonsense and depends on the ignorant credulity of creationists to even be considered a scientific argument against evolution.

And there is a particularly good example of macro-evolution which has been observed and recorded, taking place in just 36 years.

It is the Wall lizards of Pod Mrcaru

The Wall Lizards of Pod Mrcaru

In 1971 a bunch of scientists, intending to observe how a population of the Italian wall lizard, *Podarcis sicula*, adapted to an unfamiliar environment, transferred just five males and five females from the small Croatian island of Pod Kopiste in the southern Adriatic Sea, to the nearby island of Pod Mrcaru. And there they stayed while former Yugoslavia fragmented and descended into warring factions.

Thirty-six years later another group of scientists visited the island, where they discovered that not only had the teeming descendants of the relocated lizards replaced and apparently exterminated the former resident species, *Podarcis melisellensis*, on Pod Mrcaru, but that they had also diverged considerably from the original population on Pod Kopiste. That they were indeed the descendants of the original founder population was confirmed by analysis of their mitochondrial DNA, which was identical to those on Pod Kopiste.

Their findings were published in 2008 in PNAS (40).

The major significant changes were:

- A change of diet. The Pod Mrcaru lizards now eat mostly plant material, not the insects their ancestors ate.

- To cope with this different diet, the Pod Mrcaru lizards have a measurably larger head which allows for more powerful jaw muscles and a more powerful bite, needed to bite the plant matter into small chunks for swallowing and digestion.

- The Pod Mrcaru lizards are less territorial and less aggressive than their ancestors because they no longer need to defend a territory to ensure enough insects. This has enabled a much higher population density. They are also less active.

- To digest the vegetarian diet, the Pod Mrcaru lizards have developed caecal valves in their intestines. These slow down the flow of food through the digestive system and act to turn sections into fermentation vats to break down the plant cell walls.

This latter is the most dramatic morphological change since only 1% of lizard species has caecal valves. It amounts to a new structure in this species, evolved in just 36 years.

Here then we have a measurable morphological, behavioural, and ecological changes in a population in just 36 years and the evolution of new structures in the gut, all brought about by a change in the environment.

Unfortunately for creationists, the usual response to this sort of evidence of observed evolution is to fall back on an artificial distinction between what they call 'macro-evolution' and normal evolution and proclaim it to be 'still lizard kind' as though under this definition of 'kind', all lizards are the same 'kind'. However, when pressed to explain what 'macro-evolution' is, they will normally include different structures.

So, what they need to explain in this example is why evolving a new structure is not a change in kind, while fulfilling the scientific requirements for 'macro-evolution.'

Rapid Evolution in agricultural weeds

Barnyardgrass

In a research paper published in the journal Heredity in 2012 (41), three researchers explain how the discovery of agriculture by humans just a few thousand years ago has created the conditions for rapid evolution of weed plants, one of which, Barnyardgrass, *Echinocloa oryzicola* has evolved to mimic rice, making it difficult to identify and remove by hand. It has also evolved large seeds and poor dormancy making it an obligatory weed in rice paddies.

Creationists will undoubtedly wave aside the evidence of significant changes in morphology and growth habit and declare this to be the same 'kind' as its ancestral species, based on little more than the species being included in the same genus and not reclassified as a new genus.

This species is also the parent species of a new species that arose by hybridization between a tetraploid form of *E. oryzicola* and an unknown diploid *Echinocloa* species, to give the hexaploid species, *E. crus-galli*, another agricultural weed.

Waterhemp

An agricultural weed that has apparently evolved within a couple of decades is the waterhemp (*Amaranthus tuberculatus*) in corn and soybean fields in central USA.

Although *Amarantha* species are grown as grain crops, in Central and South America, this invasive weed is believed to have evolved from wild ancestors, either by *de novo* mutations or hybridization to acquire genes that enable it to tolerate herbicides. It's debateable whether this is 'macro-evolution' because the plant has acquired a new ability, and presumably metabolic pathways, which have produced a change in growth habit, or of 'micro-evolution' in response to environmental change.

In fact, the difficulty in deciding which form of evolution we should regard this as, highlights the artificiality of trying to split evolution into two types in the first place. Undoubtedly, whatever the result is, the underlying processes were the same- natural selection, and possibly hybridization and genetic drift,

Unfortunately (or fortunately for people interested in truth rather than winning arguments and fooling new recruits into joining the creationist cult) science does not have the luxury of being able to redefine its terms to suit the argument. Such is the handicap the need for intellectual integrity and the rigour of having fixed and understood definitions, imposes on science.

If we accept the scientific definition of 'macro-evolution', i.e. the evolution of a new tax on, we should include examples of the evolution of new species. And there are many recent examples of just that, some of which we can be certain evolved within a specific timeframe.

The Golden jellyfish

The following is based on a blog post I wrote in 2013 (42).

One rather beautiful example of recent evolution is that of the golden jellyfish, *Mastigias cf. papua etpisoni* that inhabits Jellyfish Lake (*Ongeim'l Tketau*) on Eil Malk Island in the tiny Micronesian state of Palau in the Pacific.

Eil Malk island is one of a group of islands known as the Rocky Islands in Palau's Southern Lagoon, the remnants of a Miocene coral reef. Jellyfish lake, like several other similar lakes, is connected to the surrounding lagoon only through the porous rock of the island.

This means that, as far as the marine environment is concerned, *Ongeim'l Tketau* is an isolated micro-environment. It has been so since 12,000 years ago when geological evidence shows was the last time the ocean level was high enough for the lake to be directly connected to the surrounding ocean.

The effect of this was to reset the clock, as far as biodiversity is concerned. At that point, every species then present in the lake became effectively isolated from its parent population and a population of (probably) the spotted jellyfish *Mastigias papua* became isolated from those in the surrounding lagoon. It mimics the sort of experiment biologists would love to do on this scale and over this timespan and is exactly the sort of experiment creationists continually claim has never been done.

The lake is one of about 200 known world-wide in which the water is stratified into distinct layers which do not mix. In Jellyfish Lake, there is a top layer which is oxygenated, and which receives sunlight, and an anoxic dark layer which is rich in hydrogen sulphide from the decaying remains on the lakebed. At the interface between these layers (known as a chemocline) there lives a group of photosynthesising purple sulphur bacteria.

Like the spotted and several other related jellyfish, golden jellyfish rely on single-celled, photosynthesising algae, which live symbiotically in cells in their 'clubs', for most of their food. The algae receive protection and are taken to the sunlight and supplied with all their nutrients by the jellyfish and supply the jellyfish with sugar in return. Juvenile jellyfish quickly build up their population of algae from the micro-organisms they take in whilst feeding in the normal jellyfish way.

What may have started off as a predator-prey relationship with the jellyfish eating the algae, or a parasite-host relationship with the algae being parasites on the jellyfish has, through the selfish interests of both genomes become a mutually cooperative and highly beneficial relationship to both, but that's not the evolution we are talking about here, though it may have progressed even further in the golden jellyfish and its algae in Jellyfish Lake. We are talking about the degree of divergence from the founder species in just 12,000 years.

In just that short time, *Mastigias cf. papua etpisoni* has undergone considerable evolution. Incidentally, the 'cf.' in the scientific name of the golden jellyfish is because it's not certain that it is a subspecies of *M. papua* and not of one of several such closely related species. *M. papua* seems the most likely candidate because it is common locally.

And this problem serves to highlight the degree of separation that a mere 12,000 years of isolation in a unique environment has produced. The changes are not just morphological either.

Golden jellyfish have unique daily pattern of migration within Jellyfish Lake.

- Night - For about 14 hours a day the jellyfish make repeated vertical excursions between the surface and the chemocline in the western basin to acquire nitrogen and other nutrients from near the chemocline for their symbiotic algae.

- From early morning to about 0930 - The jellyfish move from center of western basin to the eastern basin

- From early afternoon to about 1530 - The jellyfish move from eastern basin to near western end of lake

- As the sun sets - The jellyfish move briefly eastward from western end to western basin where they remain through the night

Spotted jellyfish also exhibit migratory behaviour in the lagoon moving with the sun as it moved across the lagoon, but it is nothing as complex as that of the golden jellyfish.

It is thought that the difference is caused by evolutionary change driven by the jellyfish-eating anemones *Entacmaea medusivora* that inhabit the eastern regions of Jellyfish Lake. The jellyfish avoid shadows and in the morning with the shadows on the eastern end the jellyfish also avoid the anemones. By moving east to west in the early afternoon the jellyfish avoid the time of day when the setting sun would eliminate shadows on the lake in the eastern end and thereby avoid the anemones in the afternoon.

So, we see not only a striking morphological change but also a change in lifestyle in as little as 12,000 years, all driven by an environmental change which first isolated a founder population and then moulded it to suit the particular micro-environment which ensued.

The East African Cichlids

Like the Golden jellyfish example, the timescale of the radiation of cichlids in the East African lakes is known with a fair degree of precision, since we know when the ecology of for creationists it was some time before their mythical 'Creation Week', it was relatively short in geological terms.

The following is based on a blog post I wrote in 2012 (43)

Towards the end of the last Ice Age, between 16,000 and 17,000 years ago there was a massive surge of icebergs and cold meltwater into the North Atlantic which caused changes in ocean currents and a climate change that produce a prolonged draught in Africa. Sediment analysis shows that Lake Victoria and Lake Tana dried up completely.

A similar event 14,000 to 15,000 years ago caused Lake Victoria to dry up again and a subsequent lowering of water levels left a small satellite lake, Lake Nabugabo isolated.

From these events we know when each lake received its founding population of cichlids from their feeder rivers. In the case of Lake Victoria, this was between 14,000 and 15,000 years ago. We also know that the micro-lake, Lake Nabugabo, has been isolated for just 5,000 years. By contrast, nearby Lake

Tanganyika is tens of millions of years old and has remained filled for all that time.

Cichlids are a group of fish which have two interesting unique characteristics:

- They have evolved a set of pharyngeal 'jaws' in their throats by fusing their lower pharyngeal bones into a single structure and a set of muscles to operate them as a secondary set of jaws, giving them the ability to exploit a wide variety of food sources.

- They are protective brooders, frequently mouth brooders: they lay a small number of comparatively large eggs which are carefully guarded, and the fry continue to receive parental protection.

These two things are believed by biologists to be responsible for the cichlids' ability to exploit new niches and so radiate into new species from a founder population.

Today, in just 14,000 - 15,000 years, Lake Victoria now has or recently had, some 500 distinct species of cichlids (44). This represents approximately twenty-two 'branching' events where each species splits into two, in other words, every branch of the diversifying tree needed to sprout a new branch twenty-two times on average, so each species needed to give rise to a new one every 650 years on average.

A series of ecological disasters have recently severely reduced Lake Victoria's cichlid population so that some 300 species are now endangered or have become extinct. These disasters include siltation as a result of deforestation and soil erosion, introduction the Nile perch and the water hyacinth and over-fishing.

I should point out that there is not full agreement in the scientific community on this timescale. Mitochondrial DNA analysis, using a hypothetical mutation rate, suggests the current diversity took between 100,000 and 200,000 years.

These two widely different timescales could be resolved in two ways: firstly, the assumptions in the hypothetical mDNA mutation rate may not be valid; secondly, there could have been a few small deep pools in which earlier populations survived the desiccation. The survival of a diverse population in small pools seems unlikely however and the analysis of the sediments seems conclusive, so on balance the shorter timescale seems the more plausible.

Nevertheless, 500 new species over even the longer timescale is impressive.

So, what creationists need to explain is how this rapid radiation into some 500 new species occurred if there has not been enough time, and macro-evolution' is impossible.

Once again, we see the dissociation between reality and creationist dogma, maintained by careful ignorance and ant-science rhetoric.

One more example of speciation should be enough, but most creationists will be well rehearsed in the standard avoidance and dismissal techniques of redefining, 'evolution' or 'species' to force fit them into 'kinds' and declare them evidence of special creation just a few thousand years ago.

London Underground Mosquito

Visitors to London should consider using a good insect repellent because there is a good chance of a new species of mosquito, *Culex pipiens f molestus*, taking a blood meal from a bear arm or leg, although their preferred source of lunch are the mice that live between the tracks.

Unlike other species of mosquito, where only the females feed on blood, it is as likely to be a male as a female that bites you, because there are few sources of the plant sap the males normally feed on in the London Underground.

Construction of the London underground started in 1863 so what we have is an example of a new species arising by population isolation in the last 161 years. During that time, the underground population has diversified from the surface population of *C. pipiens* genetically, behaviourally and in its breeding habits to the extent that some researchers now think it deserves the status of a separate species.

C. pipiens lives on bird blood whereas *C. p. f. molestus* lives on the blood of rats, mice, and humans. *C. pipiens* is cold-tolerant and overwinters by hibernating; *C. p. f. molestus* is cold-intolerant and is active all year round. The two forms do not interbreed and there is no evidence of gene flow between the two populations.

In 1999, research by Katherine Byrne and Richard A. Nichols, of Queen Mary and Westfield College, University of London, (45) showed that:

> "The surface and subterranean populations were genetically distinct, with no evidence of gene flow between closely adjacent populations of the different forms, whereas there was little differentiation between the different populations of each form. The reduced heterozygosity in the Underground populations and the allelic composition suggest that colonization of the Underground has occurred once or very few times. Breeding experiments show compatibility between the Underground populations but not with those breeding above ground."

What that means is that from a small founder population the subterranean and surface forms have diverged to the point where interbreeding is not successful (the eggs produced failed to hatch).

Because *C. pipiens* exists as *C. p. f. molestus* in other parts of its range, taxonomists regard the London Underground as another example, but in almost all respects, the London Underground is home to a home-grown new species of mosquito.

There are of course many examples of new or emerging species, a few of which I'll briefly outline, but those listed above would be regarded by most objective people as evidence that speciation is an observable phenomenon and not the impossibility that creationists would have us believe.

Other examples include:

Heliconius Butterflies

In Central and South America, *Heliconius* butterflies provide examples of speciation by hybridization (46). Hybrids of two distinct species can form stable breeding populations with distinct wing colour patterns.

These wing patterns can be naturally selected because they mimic other toxic species (Batesian mimicry). Several new species of Heliconius butterflies have evolved this way in just a few thousand years.

Apple Maggot Flies (*Rhagoletis pomonella*)

The apple maggot fly (*Rhagoletis pomonella*) provides an example of speciation in progress (47). Whether or not it has reached the status of a new species is a matter of debate, Originally, these North American flies laid eggs only in hawthorn fruits. However, around 150 years ago, a population began laying eggs in apples, an introduced plant species in North America.

The shift from hawthorn to apple trees has led to the development of distinct ecological races with different fruit preferences and breeding times. Different breeding times are a pre-zygotic barrier to hybridization which maintains genetic isolation, preventing gene flow between the two populations.

Whether or not they constitute distinct species is a matter of taxonomic classification but so far as their genetic histories are concerned, they have probably already reached that status.

New Stickleback Species

Sticklebacks, *Gasterosteus aculeatus*, though common in streams and lakes are actually a marine coastal species that frequently migrates into freshwater, finding its way upstream into lakes where it will diversify according to local conditions.

One form, the marine form, migrates back to the sea in winter then returns to freshwater in Spring. The freshwater form remains in streams and lakes throughout the year.

In various isolated lakes, these populations have undergone rapid morphological changes in traits like body size, armour plating, and feeding behaviour.

In Lake Windemere, in England's Lake District, there are two populations derived from two separate invasions: the benthic (bottom-dwelling) and limnetic (open-water) forms. These populations differ in morphology, feeding behaviour, and habitat preferences, showing signs of ecological speciation.

Benthic Sticklebacks are large-bodied, with traits suited for feeding on prey near the lakebed, such as snails and other invertebrates. Limnetic Sticklebacks are smaller-bodied, more streamlined, adapted for feeding on plankton in open water.

Transitional Forms

Creationists, probably because their understanding of evolution is the childish parodies they've been fed, think the Theory of Evolution requires a 'transitional form' in the fossil record showing how one species turned into another, as though evolution is a metamorphic process with an aim.

So, they imagine the fact that there are no half-human/half-chimpanzee or half dinosaur/half bird fossils this somehow proves evolution never happened and there is no evidence of one species evolving into another.

As though Charles Darwin is the great prophet of evolutionary biology, they like to cite a passage in 'Origins' in which he speculated that there would be transitional fossils found showing the evolutionary change in species over time.

Of course, what he wasn't saying was that his entire theory would stand or fall on the discovery of these fossils for every species, but, to a creationist with their dependence on the writings of ancient prophets, if you can prove Darwin wrong in any respect, you've destroyed the entire theory of evolution.

For anyone who understand how evolution is a gradual process that takes place in the species gene pool, not in individuals, every fossil is an intermediate between its parents and its offspring, just as you are intermediate between your parents and your children, so in that sense, not only is the claim that there are no transitional fossils wrong, but every fossil is transitional.

Having said that, however, there are a remarkable number of fossils, with more being discovered daily, that show intermediate features of diverging taxons, which is what Darwin was speculating would be found, a few of which I will describe next.

Evolution of the Giant Pterosaurs

A recent paper the journal *Current Biology* (48) showed that the flying reptile, the pterosaurs went through a ground-dwelling stage as they evolved from small, tree-climbing winged reptiles to become the predatory giants of the sky that went extinct with the dinosaurs about 66 million years ago. The team arrived at their conclusion by examining a range of fossils from about 64 taxa representing 18-20 pterosaur groups.

Descending to the ground meant their hands and feet could evolve away from their function of tree-climbing. In their abstract the authors explained:

> Small, early, long-tailed pterosaurs (non-pterodactyliforms) exhibit extreme modifications in their hand and foot proportions indicative of climbing lifestyles. By contrast, the hands, and feet of later, short-tailed pterosaurs (pterodactyliforms) typically exhibit morphologies consistent with more ground-based locomotor ecologies. These changes in proportions correlate with other modifications to pterosaur anatomy, critically, the separation along the midline of the flight membrane (cruropatagium) that linked the hindlimbs, enabling a much more effective locomotory ability on the ground.
>
> Together, these changes map a significant event in tetrapod evolution: a mid-Mesozoic colonization of terrestrial environments by short-tailed pterosaurs.
>
> This transition to predominantly ground-based locomotor ecologies did not occur as a single event coinciding with the origin of short-tailed forms but evolved independently within each of the four principal radiations: euctenochasmatians, ornithocheiroids, dsungaripteroids, and azhdarchoids. Invasion of terrestrial environments by pterosaurs facilitated the evolution of a wide range of novel feeding ecologies,

while the freedom from limitations imposed by climbing permitted an increase in body size, ultimately enabling the evolution of gigantism in multiple lineages.

The First Arthropods

The fact that insects and other arthropods have a similar body plan to the segmented worms is quite obvious from just looking at them. What is perhaps not so obvious until you look at the vertebral column is that the vertebrates are also based on the same fundamental body plan.

But what was missing in the fossil record was something showing how the segmented worms (annelids) evolved into arthropods.

But, to the disappointment of creationists who believed they had another gap in which to force-fit their putative creator god, just such a transitional fossil was found by researchers working at Durham University, UK, who have identified a species that is clearly partway between a segmented worm and an early arthropod (euarthropod).

The newly named *Youti yuanshi* is about the size of a poppy seed and fits neatly into one of creationism's beloved gaps. And, of course, it is exactly what the TOE predicts should have lived, because, as I said, the connection between segmented worms and segmented arthropods is obvious.

The Durham team, from the University's Earth Science, department, co-led by Martin R. Smith and Emma J. Long, worked with Jie Yang and Xiguang Zhang if the Institute of Palaeontology and Yunnan University, China to analyse the tiny fossil. Their findings were published in *Nature* (49).

The tiny soft-bodied Cambrian fossils were preserved by a process of phosphatization in which soft tissues of organisms are preserved by being replaced with phosphate minerals, typically apatite (calcium phosphate). This type of preservation is significant because it can capture delicate details of soft-bodied organisms that are rarely fossilized under normal conditions.

In the context of a fossil found in the Yu'anshan Formation, phosphatization means that the soft tissues of the organism, such as muscles, organs, and even cellular structures, were replaced by phosphate minerals shortly after the organism's death. This rapid mineralization helps to preserve intricate details that provide valuable insights into the anatomy and biology of early Cambrian life forms.

As well as showing primitive appendages, the fossils show the head beginning to differentiate into the various components of a typical arthropod head and an organised brain.

Tiktaalik roseae

The transitional fossil *Tiktaalik* is a major embarrassment for creationists and not just from the fact that it dates from way before their legendary 'Creation Week.

It represents a validated prediction of the Theory of Evolution which not only predicted that there would have been such a transition between the lobe-finned fish and terrestrial arthropods, but in what geological formation the fossil would most likely be found.

The missing fossil record was of the evolution from lobe-finned fish to the limbed tetrapods which were then able to colonise the land. This key step in the evolution of terrestrial vertebrates and so in the evolution of humans was estimated to have occurred about 375 million years ago, so the hypothesis to be tested was fairly straightforward - there should be 'transitional' fossils showing this stage in evolution in sedimentary rocks deposited about 375 million years ago.

So, Ted Daeschler of the Academy of Natural Sciences in Philadelphia, PA, USA and Neil Shubin from the University of Chicago, IL, USA scoured the geological maps looking for surface rocks of the right type and age and found just the right formation in the Canadian Arctic on Ellesmere Island. Work on the site was only possible during the summer so it took four summers of searching before they found what the theory said should be there. They found a fossil they called *Tiktaalik roseae* (50).

Tiktaalik, which is Inuktitut for 'burbot', a freshwater relative of the cod, had limbs which were exactly what the theory predicted, midway between the fins of the lobe-finned fish and the limbs of tetrapod amphibians with typical bones that make up a standard terrestrial vertebrate limb all easily identifiable and arranged according to the standard arrangement. And the fossil was found in rocks of the right type and the right age, just as predicted. A falsifiable prediction made by the Theory of Evolution had been confirmed.

Tiktaalik roseae and the Terrestrial Vertebrate Jaw

Not only was *T. roseae* found in the right place from the right time, as predicted by the Theory of Evolution, it also closes another gap in the fossil record – the transition from the fish jaw and method of feeding and that used by the terrestrial vertebrates.

Feeding in water normally involves suction, which is fine in water, but much less efficient in air. In air, the prey needs to be grabbed and griped before being pulled into the mouth, and *T. roseae* shows exactly that transition, showing that it spent time in both environments, leaving the water only to catch the arthropods which had already colonized the land.

The research team, which included Professor Neil H. Shubin and Edward B. Daeschler who first discovered *T. roseae*, and Dr. Justin Lemberg, from the University of Chicago, used advanced new computed tomography (CT) analysis to conduct a detailed examination of the morphology of the *T. roseae* skull (51). This allowed them to identify key new traits that had not been seen with other techniques, including sliding joints that would have allowed for the necessary cranial kinesis for the animal to expand its skull laterally to create suction.

The investigators noted distinct similarities between *T. roseae* and earlier work analysing the skulls of alligator gar, a "living fossil" fish species previously thought to only use lateral snapping motions to capture prey. In a 2019 study, Lemberg et. al (52) . found that gar use lateral snapping and suction synergistically while feeding, thanks to unique sliding joints in their skulls that help create suction while biting.

These similarities led the researchers to believe that *T. roseae* may have fed in the same way, indicating that this adaptation likely evolved long ago, before animals ever colonized land.

Transition to The Mammalian Middle Ear.

Unlike reptiles, mammals have a middle ear with three small bones called ossicles; reptile have a single bone.

The purpose of these bones is to transfer vibrations in the eardrum to the inner ear and the sensory nerves in the cochlea. It has been shown that the three-bone arrangement in mammals gives a much better sense of hearing, then that of reptiles.

But the question is, how did mammals get their three bones when their reptilian ancestors only had one? To a creationist, this is a nice little god-shaped gap so their easy answer that requires no evidence or further investigation is 'God did it!'

However, a recent discovery shows that natural selection did it without the involvement of unproven meddling of magic entities. But first, the mammalian dentition had evolved so that the double-action jaw joint of the reptiles was no longer needed, then, in a classic example of exaptation of redundant structures, two bones of the skull that had previously formed part of the reptilian jaw joint became reduced in size and migrated to the nearby ear, when they became the other two ossicles.

This had been known from previous studies of developing embryos where these forming bones could be seen to start life in the jaw joint before migrating to become part of the middle ear, so the prediction of the Theory of Evolution was that there should be an early mammal-like reptile in which this transition would be present.

And this is exactly what a team of palaeontologists, led by scientists at the Museum and the Chinese Academy of Sciences, found when they examined two small, early mammals.

The first, a mouse-sized mammal, called *Shuotheriid*, which lived between 168 and 164 million years ago, has enigmatic teeth that were previously seen as a problem for mammalian evolution. These early mammals had been placed in the same group that gave rise to the egg-laying monotreme mammals, but the researchers have shown that they were more closely related to a newly discovered species, *Feredocodon chowi*.

Analysis of the older fossil, *Dianoconodon youngi*, which dates back to between 201–184 million years ago, show that one of its two joints, the reptilian one, was starting to lose its ability to handle the forces created by chewing. The more recent specimen, *Feredocodon chowi*, already had a mammal middle ear, formed and adapted exclusively for hearing.

So, there we have a clear record of transition, firstly to a mammalian dentition which created the redundant bones that transitioned to the ossicles of the mammalian middle ear.

And all during that extraordinarily long period of the history of life on Earth that happened in the nearly 4 billion years before 'Creation Week'

Transitioning Into Insects

A central dogma of creationism, a left-over from the days when they insisted there was no such thing as evolution at all, and before they decided there were two sorts (micro-evolution, which happens all the time now and happened at warp speed after the mythical flood, and 'macro-evolution' which is still impossible) is that there are no transitional fossils. This lie is also an attempt to recruit Darwin for their superstition as he predicted there would be found series of transitional fossils showing the progress of evolution over time.

Of course, as any reading of what Darwin actually wrote, as opposed to the selective quote mines creationists are misled with, will show he never predicted there would be a complete series of transition fossils for every taxon, which would have been absurd.

This creates a problem for creationists who have to keep devising strategies for ignoring all the many examples of transitional fossils that palaeontologists keep unearthing; one of which is to claim that the transitional fossil is that of a fully formed species, which works on people who think evolution means one taxon turns into a different taxon such as cats turning into dogs or crocodiles into ducks, or, in the case of humans (the (non)evolution of whom is a special obsession with creationists), a chimpanzee turning into a fully modern human.

For a laugh, try asking a creationist to describe exactly what they would expect a transitional fossil to look like!

One such transitional fossil was discovered by a team of palaeontologists from the University of Leicester, Yunnan Key Laboratory for Palaeobiology, and the Institute of Palaeontology at Yunnan University, Chengjiang Fossil Museum, and the Natural History Museum in London.

It is that of a 520-million-year-old early arthropod, called *Kylinxia zhangi* which simply wouldn't have existed if creationist dogma had any connection with reality, as it is clearly ancestral to the insects, a group of arthropods characterises by a segmented body divides into a distinct head, thorax and abdomen, with six paired legs and two or four paired wings attached to the thorax. They also typically have compound eyes and antenna on their head and breathe through a system of spiracles.

But this fossil does not just show characteristics of insects, it also shows characteristic of crustaceans such as shrimps, crabs and lobsters, which don't have such clearly defined body-plan and typically have a cephalothorax to which legs and other appendages such as jaws are attached.

What the researchers found in a very well-preserved specimen of *Kylinxia zhangi* was a head showing the beginnings of segmentation into six distinct parts, one with eyes, the second with a jaw and the other four with legs.

As good an example of a transitional species as you could wish for and to rub salt into creationists' wounds, from 520 million years before they believe Earth was created.

A Lagerstätte of Transitional Fossils

"Sorrows come not as single spies but in battalions."

Sometimes, those creationists who have not managed to shut out all of reality must identify with Claudius in Shakespear' Hamlet, especially when science announces not just another example of a transitional species, but a veritable glut of them all from the same fossil-rich deposit or Lagerstätte.

This deposit was laid down during the Great Ordovician Biodiversification Event (GOBE) which followed the Cambrian, and a number of things contributed to environmental change which produced this rich diversification,

Plate tectonics cause the single supercontinent, Gondwana, to drift south toward the Antarctic, creating shallow seas and a shallow continental shelf. The erosion of new mountain ranges caused a surge of nutrients into the seas and an increase in oceanic oxygen allowed for more complex lifeforms.

The result was a radiation of new species of brachiopods, bryozoans, trilobites, molluscs, echinoderms, corals, graptolites, and conodonts. The latter were early jawless vertebrates with mineralised feeding parts.

In 2022, in a paper published in *Proceedings of the Royal Society B* (53), Chinese paleobiologists from the Chinese Academy of Sciences, Nanjing, China, reported on the discovery of a collection of fossils from the Lower Ordovician showing the transition between the Cambrian and Ordovician biota forms from the early part of this radiation, which reached a peak between 460 and 470 million years ago.

They reported that the collection "contains a variety of soft tissues, as well as rich shelly fossils, including palaeoscolecidan worms, possible *Ottoia*, trilobites, echinoderms, sponges, graptolites, polychaetes, bryozoans, conodonts, and other fossils. The fauna includes taxa that are not only Cambrian relics, but also taxa originated during the Ordovician, constituting a complex and complete marine ecosystem. The coexistence of the Cambrian relics and Ordovician taxa reveals the critical transition between the Cambrian and Palaeozoic evolutionary faunas.

A Feathered Pterosaur

It's just about universally accepted nowadays except in reality-denying creationist circles that birds evolved out of feathered dinosaurs, but the question was why did dinosaurs evolve feathers.

Feathers in birds serve two, and possibly three functions in addition to the flight feathers that form the control surface of wings and tail; firstly they insulate, secondly they provide colour which is important in displays, both territorial and attracting a mate and for camouflage and thirdly, they help control airflow over the body and particularly over the wing surface, reducing drag and increasing lift.

But what would a dinosaur use them for?

The obvious thing is for insulation, and maybe sexual displays the way ostriches use them. Now research by an international team of palaeontologists led by Dr Aude Cincotta and Prof. Maria McNamara from University College Cork (UCC), Ireland and Dr Pascal Godefroit from the Royal Belgian Institute of Natural Sciences, together with colleagues from Belgium and Brazil, have discovered a 115-million-year-old pterosaur, *Tupandactylus imperator* from Brazil, with a pronounced, feathered head crest.

The feathers are short, wiry feathers and fluffy, branched feathers, just like those of birds, along the base of the crest. Viewed under a powerful electron microscope, the specimen had preserved melanosomes – granules of the pigment melanin. Unexpectedly, the new study shows that the melanosomes in different feather types have different shapes, showing that the different feathers had different colours.

The team concluded: "These tissue-specific melanosome geometries in pterosaurs indicate that manipulation of feather colour—and thus functions of feathers in visual communication—has deep evolutionary origins. These features show that genetic regulation of melanosome chemistry and shape was active early in feather evolution."

In other words, feathers were already transitioning into the feathers of today's birds and serving some of the same functions, 115 million years ago.

Transitional Crabs

True crabs are found all around the world, from the depths of the oceans to coral reefs, beaches, rivers, caves, and even in trees as true crabs are among the few animal groups that have conquered land and freshwater multiple times.

The crab fossil record extends back into the early Jurassic, more than 200 million years ago. Unfortunately, fossils of nonmarine crabs have been sparse and restricted to bits and pieces of the animal's carapace – claws and legs found in sedimentary rocks.

That is until now with the discovery of an exquisitely preserved specimen of *Cretapsara athanata*. The crab was discovered in 100-99-million-year-old amber from Myanmar. It has been examined by a team of palaeontologists led by Lida Xing, of the China University of Geosciences, Beijing, and including Dr Javier Luque, of the Department of Organismic and Evolutionary Biology, Harvard University, USA, using a CT scanner which created a full three-dimensional reconstruction of the crab. This allowed Luque, Xing, and their team to see the complete body of the animal including delicate tissues, like the antennae and mouthparts lined with fine hairs.

Surprisingly, they discovered the animal also had gills.

The team have published their findings, open access, in *Science Advances* (54).

Marine crabs have gills and terrestrial crabs have lungs, so what was a crab with gills doing getting trapped in amber?

Clearly, the crab was transitional between a marine and a terrestrial crab, and we have another of those transitional forms creationist counter-factual dogma insists do not exist.

Just a couple more examples of transitional fossils should suffice, then I'll move on the something creationists go into paroxysms of denial over – the evidence of human transition from a common ancestor with the other African great apes particularly the chimpanzees, through the Australopithecines and archaic hominins to modern *Homo sapiens*.

The Panda's Transitional 'Thumb'

Pandas are an evolutionary oddity. They are bears of the Carnivora order of mammals, yet they live on a diet of almost exclusively bamboo.

To facilitate holding the bamboo while they chew it, they have evolved the so-called Panda's thumb which acts as a sixth digit and opposable thumb. This extra digit has evolved out of the sesamoid bone of the wrist and so must serve two functions- holding bamboo and weight-bearing while walking.

These two requirements are opposite forces in evolution, as can be seen in the fossils of an ancestral panda, the *Ailuropoda melanoleuca*. This species lived about 6-7 million years ago, by which time it had already evolved the extra digit, with a difference. Their extra digit was long and straight, unlike the shorter, stouter, and slightly hooked bone of modern pandas (55).

It would have functioned perfectly as a thumb for gripping bamboo but not so well for walking.

This illustrates how an evolutionary process can provide a crude quick fix for a need. In this case the advantage of being able to grip bamboo shoots while eating them far outweighed any slight reduction in walking efficiency. But when the ability to grip bamboo had evolved, the problem of walking efficiently produced selection pressure which was no longer over-ridden by a greater need, so, as long as there was no loss in the ability to grip bamboo, evolution could improve the wrist bone's other function of weight-bearing.

The result is the improved version of the panda's 'thumb'. The crude but functional extra digit of the *Ailuropoda melanoleuca* was the panda's thumb in transition.

The Spider's Tail

Spiders, a group of arachnids are characterised by modified body appendages called spinnerets which extrude silk, and, in males, a pair of pedipalps which are used to insert sperm into females. All but the most primitive spiders also have smooth, non-segmented abdomens. Scorpions, on the other hand have segmented abdomens and lack spinnerets and pedipalps. They also have the characteristic tail.

The two groups are believed to have diverged about 430 million years ago but some years ago palaeobiologists discovered a fossil arachnid, *Chimerarachne yingi* gen. et sp. nov. that was 100 million years older than the earliest spiders, but which had spinnerets. These were put in an archaic group, the Uraraneida, relatives of the spiders but not necessarily their direct ancestors.

Then, in 2018, two teams of palaeontologists examined a specimen of *Chimerarachne yingi* embedded in Myanmar amber and concluded that it was indeed a Uraraneida. One team examined details of the dorsal surface: the other the ventral surface. They published their papers in the same edition of *Nature, Ecology & Evolution* (56) (57).

In effect, *Chimerarachne yingi* is a spider with a tail that existed alongside true spiders for some 100 million years. It had well-developed spinnerets, spider-like chelicerae and the males have pedipalps. But it had a tail, the vestige of the common ancestor it shared with scorpions.

What this shows is that the early arachnid ancestors of spiders and scorpions had a mix of the characteristics of both, and that evolutionary divergence consisted of loss of some characteristics and enhancement of others. Spiders lost their tails and segmentation of their abdomen and scorpions lost their spinnerets and pedipalps.

Here then we have a clear pointer to a past when the common ancestor of spiders, scorpions and these Uraraneida had intermediate characteristics between the two present day groups, spiders and scorpions and a clear evolutionary explanation for these beautiful fossils as well as the modern groups.

Would that be enough to convince a creationist that their cults dogmatic rejection of intermediate or transitional forms and denies they exist in the fossil records?

How well is their personal 'Morton's Demon' working?

Maybe examples of human ancestors with transitional features will do the trick. I'm not taking any bets.

Transitional Hominins

If you want some fun, as a creationist who has just declared that there are no transitional fossils, what they would expect one to look like. It is a rule of the cult that they must never give a definition of anything they believe does not exist; in case someone produces one.

So, in the absence of a definition of a transitional fossil that a creationist would accept as one, I will propose a simple definition:

A transitional fossil will have a mixture of the features of its ancestor and what it transitioned into.

So, for example, a transitional fossil showing that Australopithecines transitioned into hominins, would have a mixture of Australopithecine features and hominin features.

Australopithecus sediba

In August 2008, Matthew Berger, the son of palaeoanthropologist Lee Berger, found the fragment of a human-like collar bone. The first excavations at the discovery site, Malapa, north of Johannesburg, were performed by a team from the Swiss Field School of the Anthropological Institute at the University of Zurich, under the direction of Peter Schmid.

The fossils did not match any previously known hominin species. Based on age and morphology, researchers carefully allocated the new hominin species to the *Australopithecus* genus and not to the *Homo* genus and named *Australopithecus sediba*, which in the Sesetho language means "fountain" or "source." (58)

In 2011 a team based at the University of Zurich concluded that, *Au. Sediba* could be directly ancestral to the *Homo* genus in five papers in the journal *Science* based on the observation that "*Australopithecus sediba* unites various properties that were not yet seen in early ancestors of humans. The fossils show a surprisingly modern, yet small brain; a very modern, developed hand with long thumbs, like in humans; a very human-like pelvis; but a form in the foot and ankle shape that is both ape- and human-like. In light of these findings, Prof. Lee Berger, University of Witwatersrand, is of the opinion that *Au. sediba* is the best candidate ancestor for the genus *Homo*

On of the investigators, Peter Schmid from the University of Zurich, wo co-author the publications, noted that, based on the cranial capacity of a 10-13-year old child that an adult *Au sediba* would have had a cranial capacity of about 440 cm^3 which is within the 282-454 cm^3 range of a modern chimpanzee, despite the very hominin-like face and locomotory system.

Apart from the foot and ankle which has a mixture of ape and human characteristic, *Au. sediba* looks like a human torso with a chimpanzee-sized brain, although the cranium shows signs that the frontal lobes of the brain were more advanced than those of a chimpanzee.

The hands, in particular look even more adapted for toolmaking than those of *Homo habilis* (Handy man).

In short, *Au. sediba* had a mosaic of Australopithecine and *Homo* features (59) that some authorities believe means it should be reclassified as *Homo sediba* and regarded as the 'missing link' between the Australopithecines and the Hominins.

On an amusing side note, I once showed an ardent creationist who insisted that there were no transitional fossils show humans had evolved from earlier ancestors, photos of a modern human skull, the skulls of a chimpanzee and a gorilla and the skull of *Au sediba* and asked him which the *Au. sediba* should be regarded as. "It's an extinct ape!" he declared. I then showed him photos of the hands of the same species and asked the same question. "It's a human hand!" he declared, saying he could also tell it was a manual worker.

I then showed him a photo of the complete skeleton of Au. sediba with the skull and hands he had identified as Ape and human, respectively and asked him why a species with an ape skull and a human hand should not be regarded as a transitional fossil. He promptly broke of communication and left the social network group. I assume he is still chanting his protective mantra, "There are no transitional fossils!" in other social media groups.

In an even more bizarre instance, I did the same with another creationist, to be told "I don't care whose hands it has, *Australopithecus* is an extinct Australian [sic] ape!" To howls of laughter, he quickly left the group and soon deleted his account.

But no matter that you show creationists an example of exactly what they claim there are no examples of, you can expect a swift moving of the goal posts. For example, "hominins are all the same 'kind'! what about the 'missing link' between humans and chimpanzees?

So, I give you:

Sahelanthropus tchadensis

Otherwise known as the Sahel ape from Chad, a fertile area immediately south of the Sahara and, in more recent human history, believed to be an important migration route for human groups moving between the Nile valley and West Africa.

In 2013, an article appears in Scientific American confirming that a seven-million-year-old fossil skull, nicknamed Toumaï, found in 2002 in Djurab Desert in Chad, Africa, may well be the oldest known ancestor of *Homo sapiens*. The species the skull was from had been given the name *Sahelanthropus tchadensis*,

The discoverers, a team lead by Michel Brunet, a palaeontologist at the University of Poitiers, France, had always claimed that the skull was from a species close to the point of departure of *Homo* and *Pan* (Chimpanzees) but the question was on which branch of the diverging tree it should be placed - in other words was it the skull of a hominin or an ape.

Now Thibaut Bienvenu of the Collège de France and his colleagues have managed to reconstruct the endocast of the inside of the brain case and so infer the shape of the brain which once occupied it. They did this by imaging it with 3D X-ray synchrotron microtomography, which is a technique based on high-energy x-rays produced by electrons accelerated in a synchrocyclotron, which have enough power to smash through hard materials such as the mineral matrix which filled the interior of the skull.

This technique produces a computer image which when processed to remove the matrix and allow for deformity of the skull, showed unmistakeable signs of a hominin brain the size of that of a chimpanzee:

In a presentation given on April 2, 2013, at the annual meeting of the Palaeoanthropology Society, Bienvenu reported that the endocast shows strongly posteriorly projecting occipital lobes, a tilted brain stem and a laterally expanded prefrontal cortex, among other hominin brain characteristics.

Other hominin-like features are relatively small canine teeth, associated with reduced aggression and the forward position of its foramen magnum (the spinal cord opening in the base of the skull), which is associated with upright walking.

Transitional Forms

The difficulty in deciding whether this is a *Pan* or a *Homo* is precisely what we would expect of an early human ancestor close to the divergence of humans from chimpanzees. We would expect it to have characteristics of both and characteristics such as the heavy brow ridges which have been retained in chimps, are present in many other early hominids but which are absent in most humans today.

As we move back in time towards when the distinction between any two diverging species was blurred, so we expect the difficulty in distinguishing between them to increase and revolve around finer points of detail, and with that, the arguments to be less easy to resolve.

This is science.

Australopithecus afarensis (Lucy)

About 3 million years ago a creature midway between a chimpanzee-like ape and a human being fell out of a tree and died of her injuries.

This creature was the best known of some 300 specimens of a species now known to science as *Australopithecus afarensis*, a species which is one of the candidates for being the direct ancestor of the *Homo* genus.

She is known to the world as 'Lucy'.

Lucy was about 3.5 feet tall, weighed about 60 pounds and, judging by her curved fingers, was a member of a species which probably still foraged in trees or took to them for safety. But she was also a member of a species which, judging by her pelvic girdle, lower limb and feet, spent a great deal of their time on the ground walking upright. The latter would have compromised her arboreal dexterity having lost the ability to grip effectively with her toes, unlike her more chimpanzee-like ancestors.

Now scientists believe they have pieced together her last few seconds of life as she fell from some 40 feet, probably out of a tree, hitting the ground at about 35 miles an hour, landing on her feet and falling forward onto outstretched arms.

The fractures in her arms, legs, ribs and skull all point to being perimortal. The investigation was carried out by a team led by John Kappelman, an anthropologist at the University of Texas at Austin, using a powerful form of CT scanning known as High-Resolution X-ray Computed Tomography Facility. (60)

Au. afarensis fossils have been found in and around the Afar region of Ethiopia, and further south in Southern Africa. The same species is believed to have been responsible for the footprints left in fresh volcanic ash at Laetoli (61) before it became solidified, probably by rain. These footprints show the species that made them was fully bipedal and had the lower limbs and feet of a hominin, so again we have the familiar mosaic of hominin and Australopithecine features of a transitional species.

Homo naledi

Another candidate for the 'missing link' between Australopithecines and the Hominins is a species that was discovered in the Rising Star Cave system, Gauteng province, South Africa. It lived about 335,000–236,000 years ago.

There is some controversy over whether the remains were deliberately and ritually placed in the cave – which would imply some sort of religious belief or belief in an after-life – or whether they were simply disposed of in the inner chamber in the cave system, but what is not in doubt is that they again have a mosaic of Australopithecine and Hominin features although they have been placed in the Hominin clade.

They had a smaller cranial capacity than later Hominins but larger than that of Australopithecines (465–610 cm^3). Modern humans have a cranial capacity of 1,270–1,330 cm^3.

They averaged about 5'9" tall and had a 'cephalization index' close to that of other contemporary Hominins so probably had similar cognitive abilities.

But they are enigmatic in that they have a relatively small brain at a time when other African Hominins had evolved larger brains and were evolving towards that of *Homo sapiens*. Why they retained that low brain size when the same environmental conditions were causing other Hominins to evolve much larger brains is yet to be determined.

Evolution of the Balance Organ for Bipedalism.

Palaeontologists and palaeoanthropologists can generally tell the differences between different taxons of humans by measuring and observing the details of bones and teeth, but what is not always obvious are the more subtle changes that occurred during human evolution, such as the changes in the structure of our inner ear and especially the balance organ which was essential both for swinging around in trees and walking on the ground. And it is this part of ours and our ancestor's anatomy that palaeontologists has shown a clear transitional series from the mode of locomotion of early apes, through the great apes and to humans.

To walk upright successfully needs a fully functioning balance organ in the inner ear, as anyone suffering from Ménière's disease will testify, so the study of the origins of bipedalism in the remote ancestors of humans needs to consider changes in the inner ear that would facilitate it.

Humans and our closest relatives, the great apes and the simians, display a range of locomotion but only humans are normally fully bipedal, although chimpanzees can use bipedal locomotion when carrying a load for example.

The monkeys normally run along branches on all fours, balanced on top of them and jumping from branch to branch; the apes hang beneath the branches in locomotion known as brachiation, but humans are ungainly in trees and prefer bipedal locomotion on the ground. The question is, when did this ability evolve in our ancestry?

We can be sure our hominin ancestors the Australopithecines, were fully bipedal because we have a record of their footprints in volcanic ash at Laetoli, and their lower limbs and feet were almost indistinguishable from those of Homo sapiens. 'Lucy' (*Au afarensis*) was probably mostly bipedal but may have taken to trees for safety and possibly to sleep on constructed platforms like chimpanzees do. The evidence of injuries to her fossilised skeletal remains suggests she may have died by falling out of a tree.

To investigate this stage in our evolution a group of researchers, led by Professor Xijun Ni, which included Yinan Zhang, a doctoral student, both of the Institute of Vertebrate Palaeontology and Palaeoanthropology of the Chinese Academy of Sciences (IVPP), and Terry Harrison, a New York University anthropologist, used 3-dimensional CT scanning to examine the inner ear of a 6-million-year-old fossil ape, *Lufengpithecus*, unearthed in China's Yunnan Province in the early 1980s, and compared it to the inner ear of other living and fossil apes and humans from Asia, Europe, and Africa.

The formation the fossil was found in has been previously dated magnetobiostratigraphically to about 6 million years. This technique depends on the record of periodic changes in Earth's polarity trapped in magnetic particles in sedimentary rocks and by recording the microfossils such as pollen associated with these changes:

The team have published their work, open access, in the Cell Press journal, *The Innovation*. Their results show that the split between the gibbons and the other apes was reflected in changes in the semicircular canals and again the split between chimpanzees and humans involved another change in this balance organ, so we have another transitional series, this time to the detailed structure of our ancestors' middle ear.

Transitional Forms and Creationist Duplicity.

Creationists confusion over 'missing links', 'transitional forms', etc, comes from their childish parody version of evolution that they seem to be determined not to be shifted from. They demand to see a 'transitional form because they believe evolution is one individual changing into another or at least giving birth to an offspring halfway to becoming another species. And there is always an assumption that evolution has a goal.

One 'Gotcha!' question I was asked recently, as a diversionary tactic, was what were the parents of the first human? As though there ever was a first human when modern humans evolved in the gene pool of an earlier species of *Homo*, which ultimately evolved out of the gene pool of a species of *Australopithecus*. There never was a first of any species in an evolving and slowly changing continuum. All we can show, because that is all that can be expected from the way evolution works is a fossil showing intermediate features from somewhere along that continuum. I often use the analogy of a rainbow where violet changes into red via a number of 'intermediate' colours, but nowhere in that continuum is it possible to say exactly where one colour turns into another.

The dishonest and fraudulent nature of creationism and the childish ignorance of creationists is exposed by the continual demands that science provides evidence of something that never happens, and that science does not claim happened. Science can always provide evidence of things it claims happened because it only ever makes those claims if there is evidence, or strong logical argument for it.

For example, we can be sure that there were archaic humans living in Africa a few million years ago and that their descendants gradually became modern humans. We can also be sure that their populations split and diverged at times into other species that then either became extinct or were absorbed into another population of humans. We know this because we have snapshot fossils from various times and places that show that sequence of change over time.

What we don't claim is that a primitive hominin ever gave birth to a more advanced hominin, or that a population of hominins got together and decided to evolve a bigger brain or the ability to walk upright. And what we do not claim is that every generation must have deposited a representative fossil somewhere where we can easily find it.

It would take someone with the mind of a child and a determination to stay ignorant of what they are attacking to suppose that science ever makes those absurd claims., or even more absurdly, that a failure to prove what it doesn't claim is a failure of science that renders the entire body of science worthless and unreliable.

Something From Nothing

There are two points in the development of the Universe and life on Earth where creationists think their teleological arguments are unanswerable, because the only answer they can imagine is 'God did it! And it goes without saying, the god who did it was the locally popular one their parents told them to believe in. Those two points are:

- The Big Bang.
- Abiogenesis.

I'll take each in turn and show how there are plausible scientific explanations for them, and plausibility is enough to refute the claim that there are no possible natural processes, so the only answer is, 'God did it!'

The Big Bang.

The Big Bang is the singularly inappropriate name given to the instant at the beginning of time and space that is predicted by Einstein's Relativity and confirmed by the Red Shift and the Background Microwave Radiation. The term was coined by the astronomer, Fred Hoyle with the intension of mocking the theory that ran counter to his 'Steady State' theory. It was neither big, nor did it go bang. It was a very small, silent event because without matter there is no way sound can be produced. But. somehow the Little Silence does not quite conjure up the drama about to unfold.

It was the point in time (t) where t=0 and 'everything' was compressed into a super-dense, infinitely small point. Except that is not quite true, because most of what now exists didn't. What existed was a quantum zero with all the potential of an unbounded fluctuation around zero.

Although the 'singularity' which existed when t=0 is a prediction of Relativity, below an, as yet, undefined level, the predominant domain is that of quantum mechanics where the laws of Relativity break down and quantum effects take over.

As a quantum-level event, there is no need to speculate on a cause because causality is a phenomenon of Relativity; at the quantum level uncaused events are commonplace, and causality when t=0 is meaningless because cause needs prior time so one event can follow another in time.

Examples of uncaused quantum events are:

- The spontaneous generation of particles in a zero-energy quantum field, as seen in the Casimir effect (62).
- The decay of an excited orbital electron to its ground state
- The decay of radioactive isotope.

Another effect of quantum mechanics is that it imposes a constraint on the minimum length of time that can exist. This is known as Planck Time (t_p) which is derived from three fundamental constants:

1. Gravitational Constant (G): Describes the strength of gravity.
2. Speed of Light (c): Sets the maximum speed of information transfer and is fundamental in relativity.
3. Planck's Constant ($\hbar$): Governs quantum mechanical effects, describing the scale at which quantum behaviour dominates.

So, by combining these constants, we can say:

$$t_p = \sqrt{\frac{\hbar G}{c^5}}$$

Which approximates to 5.39×10^{-44} seconds. In other words, blink and you have missed trillions of them. Notice that the three constants are all constants within Relativity, so in effect, what this formula means is that's the smallest unit of time to which Relativity applies; below that we enter the strange realm of Quantum Mechanics with which our intuition is singularly ill-equipped to cope, because it evolved in the 'Macro' world of Relativity.

What the tiny amount of time, which is the minimum age the Universe could be, gives to quantum mechanics is enough time for three of the four fundamental forces to split from the fourth, gravity, and produce a super-dense hyperinflationary increase in energy and the rest his history – literally. An entire universe not from nothing as creationists believe but from an uncaused fluctuation in a non-zero quantum energy field.

Curiously, creationists whose simplistic answer is that a god made of nothing made everything from nothing with some magic words, always accuse 'evolutionists' of believing the Universe came from nothing by going bang.

Incidentally, because there are four fundamental forces each a form of energy, and because three of them, the weak and strong nuclear forces and electromagnetism added together equal the fourth, gravity, which works in opposition to the other three we can say the total energy in the Universe is zero.

The simple maths is:

$$+\text{Lots} - \text{Lots} = 0$$

Which mean in energy terms, the Universe is nothing!

A good analogy is that of a bank loan. The bank lends you $1,000, now the bank has -$1,000 and you have +$1,000. Together there is $0 but the bank has an asset (your debt) and so do you. No wealth was created in that transaction.

Abiogenesis

Creationists have a strange view of abiogenesis and in particular what 'life' is, so they constantly ask evolutionary biologists to explain how 'life' can come from 'non-life' – something which they believe is impossible, because their cult dogma says so. But try to get into a discussion about what they imagine 'life; is and how it can be detected, and all you will get is evasions. It is never clear whether they think it's a substance, a process, a force or something magical, probably because they've never g. n it much thought themselves.

In essence, 'life' is what we call the processes by which an organism maintains itself in its environment and resists the tendency to increased entropy, by utilising the energy in the nutrients it consumes.

In other words, life is entropy management.

The other objection creationists have to the concept of natural abiogenesis, is that a complex cell is too complex to have simply arisen by random chance out of inorganic chemicals.

In other words, their objection is not to the chemistry and physics of how a living process could have arisen through an evolutionary process, but to an assumed vastly small probability of their parody of abiogenesis.

Of course, no serious biologists believe a complex cell arose fully formed by random chance, complete with all its organelles, and there are several plausible explanations by which inorganic chemicals, and maybe some precursor molecules such as amino acids and nucleic acids could have formed simple metabolising systems enclosed in a lipid membrane and then evolved over time into a simple, free-living, self-replicating cell.

The following is from my book, *What Makes You So Special: From the Big Bang to You* (63). It was based on a 2009 article in New Scientist by Nick Lane and Michael Le Page (64).

Something From Nothing

1. Water filtering down into newly formed rocks around geothermal vents reacted with minerals to produce an alkaline, hydrogen and sulphide rich fluid that welled up in the vents.

2. This fluid reacted with acidic sea water which was then rich in iron to form deposits of highly porous carbonate rock and a foam of iron–sulphur bubbles.

3. Hydrogen and carbon dioxide trapped in these bubbles reacted to make simple organic molecules such as methane, formates and acetates; reactions that would have been catalysed by iron–sulphur compounds.

4. The electrochemical gradient between the alkaline fluid in the pores and the acidic seawater would have provided energy to drive the spontaneous formation of acetyl phosphate and pyrophosphate. These behave like ATP (adenosine triphosphate) which powers modern cells. This power supply would in turn power the formation of amino acids and nucleotides.

5. Currents produced by thermal gradients and diffusion within the porous carbonate rock would have concentrated the larger molecules creating the conditions for building RNA, DNA and proteins and creating the conditions for an evolutionary process where molecules that could catalyse the formation of copies of themselves would quickly dominate and win the struggle for resources.

6. Fatty molecules would have coated the surface of the pores in the rock, enclosing the self–replicating molecules in a primitive cell membrane.

7. Eventually, the formation of the protein catalyst, pyrophosphatase enabled the protocell to extract more

energy from the acid–alkaline gradient. This enzyme is still found in some bacteria and archaea.

8. Some protocells would have started using ATP as their primary energy source, especially with the formation of the enzyme ATP synthase. This enzyme is common to all life today.

9. Protocells in locations where the electrochemical gradient was weak could have generated their own gradient by pumping protons across their membrane using the energy released by the reaction between hydrogen and carbon dioxide, so producing a sufficient gradient to power the formation of ATP.

10. The ability to generate their own chemical gradient freed these protocells from dependence on the pores in the rock, so they were now free to become free–living cells. This could have happened at least twice with slightly different cells, one type giving rise to bacteria; the other to archaea.

The above ten–step process is of course speculative and impossible to test and verify in a laboratory because the conditions around these geothermal vents deep below the ocean would be impossible to replicate in a laboratory, as would the time it might have taken. No–one is claiming it all happened in a day or two, or even weeks or years, not even the lifetime of a working scientist. It could have taken tens or hundreds of millions of years. No–one was in any hurry and there was no objective. Things happened when they happened.

But the point is that this is a very plausible mechanism for producing the first prokaryote cells. We do not know for sure what the actual process was, and we probably never will, but one thing is sure: it can be done without magic and without a special ingredient called 'life'.

The Genetic Code

Like other creationists objections this one is based on a childish parody of what scientists think really happened. The trick of their cult leaders is to get them to think that scientists believe something so stupid that they must all be mad. This makes creationist feel superior to scientists.

The objection is another of those vanishingly small probability arguments based on the false claim that scientists believe an entire species genome self-assembled spontaneously as a single event, with every one of billions of bases in precisely the right order, simply by random chance.

At its basics, this is the same as the card-dealing thought experiment I used earlier in the chapter on the misuse of statistics. It assumes the outcome was the intended outcome.

But the Theory of Evolution By Natural Selection is quite capable of explaining how an enzyme coded for by RNA can be improved gradually over time by the more successful variants being retained and increasing in the gene pool while less efficient mutations are eliminated. And, as I explained earlier, new genetic information can be acquired in a number of different ways.

But having explained that to a creationist, there will almost certainly be the traditional moving of the goal posts and the demand will be to explain how the 'genetic code' arose by chance.

The generic code is of course the triplet code whereby triplet codons code for specific amino acids, so RNA acts as the template for building amino acid sequences into proteins, each of which has a particular function – as enzymes, structural proteins, etc.

What you will never do is get a creationist to explain how many different genetic codes there could have been so we can estimate the probability of just the present one arising. It is the 'fine-tuned' fallacy all over again where from a sample size of one, creationists pretend they can calculate the probability of a particular triplet code and claim it must be vanishingly small. But if there are no other possibilities, the probability is not vanishingly small but certainty.

However, there are good reasons to think there is a natural evolutionary explanation for the triplet code being as it is and being an evolutionary process, it improved over time from simple beginnings, one such improvement being the minimization of errors.

For those unfamiliar with the triplet 'code' (I use the term code because that's how it's popularly known, but that leads people down the path of assuming it must be like computer code, which, with their traditional argument from ignorant incredulity and false dichotomy fallacies – "I don't understand this, therefore God did it – creationist declare that a computer program requires a programmer, i.e., the locally popular god their parents told them about. In fact, a better analogy is that of a template).

There are four bases in DNA, thymine (T), cytosine (C) adenine (A) and Guanine (G). These can occur in any order to give $4^3 = 64$ different three-letter sequences: ACT, AGT, ATG, etc.

However, there are only about twenty amino acids, so, even allowing for some triplets to code for things like 'stop' or 'start' there is plenty of scope for redundancy.

When DNA gets translated to RNA, Thymine is replaced by the chemically very similar Uracil, (U) otherwise the triplet codes are the same.

Work is still ongoing in this area of research, but some current theories are:

The Error Minimization Hypothesis,

This hypothesises that the triplet code evolved to reduce the impact of mutations or errors in protein synthesis. Codons encoding similar amino acids often differ by only one base, which helps prevent drastic changes in protein function when single-base errors occur.

This is illustrated by how single-base mutations in codons tend to swap amino acids with similar chemical properties. For example, if the codon for glycine (GGA) mutates to GGU or GGC, it still codes for glycine, maintaining protein stability. This built-in redundancy makes the genetic code robust, protecting against harmful mutations.

In this instance, a random mutation in the triplet code has a one in three chance of not making any difference, which is a significant advantage over any mutation resulting in the wrong amino acid in the protein chain. However, with some amino acids being chemically similar to others, even that mutation might not be disastrous.

The Stereochemical Hypothesis

This idea proposes that, in early evolution, certain RNA molecules had specific chemical affinities for particular amino acids. For instance, some experiments have shown that codons or anticodons for alanine (GCU, GCC) may have a weak chemical attraction to alanine itself. Although not universally supported, these findings hint at a primitive "language" where certain codons directly interacted with amino acids, creating early associations between nucleotides and amino acids that might have set a foundation for the code.

The Co-Evolution Theory

This idea builds on the Stereochemical Hypothesis by suggesting that the code expanded in tandem with the availability of new amino acids. Primitive life may have initially used a simpler code, gradually incorporating more complex amino acids as they became accessible through metabolic innovations.

For example, consider that simpler amino acids, like glycine and alanine, are likely ancient and might have been part of an early, simpler code. Over time, as biochemical pathways evolved, organisms began synthesizing more complex amino acids like tryptophan. These new amino acids found their place in the genetic code through additional codons, expanding the code's complexity. This gradual evolution suggests the code may have started with fewer amino acids and expanded as more amino acids became available through evolution. In that case a simple two-letter code which coded for 16 amino acids could have been enough for the number of amino acids plus some other codes like 'stop' and 'start', with the third letter being added later because it improved stability.

These ideas can be combined with:

The Frozen Accident Theory

Which posits that once established, the code became "locked in" due to the high cost of genetic code changes—these ideas illustrate a multifaceted evolutionary process.

Together, these theories suggest that a mix of selective pressures, biochemical constraints, and historical accidents shaped the code. This multi-factor approach aligns with the near-universal use of the genetic code across life forms, reflecting its fundamental role in the evolution and stability of life.

So, rather than the creationist false dichotomy of either a completely random, highly unlikely event with a vanishingly small probability, or the locally popular god did it, there is a highly probable evolutionary explanation for the genetic code.

Something From Nothing

How Creationists Mislead Their Supporters

It might seem strange that the leaders of the creation cult who are all religious fundamentalists, will knowingly misrepresent science and scientists. After all, they are pushing the line that the Bible is the inerrant word of God who will one day come to judge them on how well or badly they've stuck to its instructions (and I include Moslems in that category because the Bible is regarded as holy book by Islam). One of, the instruction, supposedly from that God, is that they must not bear false witness. In other words, they shouldn't lie, especially about other people.

And yet the creationist literature is full of misquotes, quotes taken out of context and presented as saying the opposite of what the author said, blatant misrepresentations of science and accusations, direct or implied that scientist lie and falsify their results. Accusations that would be more accurate if directed at the same creationists making them.

For example:

Radio-halos

In a book, *It's a Young World After All* (65), by a notorious creationist, Dr Paul D. Ackerman, PhD. (whose PhD is in psychology, not biology, geology or physics) he cites the findings of a fellow creationist, Robert V. Gentry, who claimed to have found evidence of 'radio-halos' in coal from a coal measure in Colorado, USA, that show the coal is only a few thousand years old. Radio-halos are frequently cited by creationist as evidence of a young earth.

Ackerman presents Gentry's conclusion as established scientific fact.

Briefly, radio-halos are caused by the decay of radioactive isotopes of uranium through a number of intermediate isotopes of other elements, to a stable isotope of lead, as I described in the earlier chanter on geochronology. Each of the intermediate isotopes leaves a characteristic ring around the original radioactive atom creating the three-dimensional 'halo'. The distance from the emitting particle depends on the energy of the radiation, so is characteristic of each decay in the series.

The scientific approach would be to conclude that, with some many other measures showing Earth is billions of years old, there is probably a reason why 'radio-halos' are giving a result so far out of line with all the others. The creationist fraud's approach is to claim that one anomalous result proves all the others are wrong and probably falsified. Confirmation bias and pandering to the conspiracy theory of science.

Of course, there is an underlying assumption which Ackerman fails to point out, that the uranium got into the sample of coal when it was first laid down, although there is no reason to think that is true. And in fact, this assumption is false because coal is not watertight so uranium and other radioactive isotopes can leach in from the surrounding soil. at any time after it was laid down.

What Ackerman also failed to point out was that Gentry's claim has been comprehensively refuted and is dismissed by geologists as bogus (he is a creationist after all and would not have been permitted to publish his conclusions in creationist literature had it not been in full accord with a literal interpretation of the Bible. (66)

Gentry's finding was refuted by J. Richard Wakefield (67), who pointed out the Gentry had not carried out any investigations of the surrounding geology to find possible sources of contamination.

There are also a number of other errors in Gentry's methodology. He claimed, for example that the halos were caused by polonium, which is one of the products of uranium decay, but, since there was no uranium present, and polonium has a relatively short half-life, the polonium must have been included in the original 'creation' of the coal. But, as Wakefield pointed out, it is not possible to distinguish between the haloes of polonium-210 and those of radon-222 (another of the decay products of uranium), so the halos Gentry was making much of could well have been produce by the radioactive gas, radon-222 which would have been free to migrate within the coal and not stay at the center of decay.

And nor are those the only problems with Gentry's methods, so his conclusions can't be taken seriously, yet the readers of Ackerman's creationist polemic are left with the impression that Gentry has provided definitive proof of a young Earth. It would have taken about 5 minutes of fact-checking for Ackerman or his publishers to have discovered that this was a false claim.

Ackerman's book is a classic of creationist misrepresentations of the science about which I wrote a series of refutations in my blog. (68). Perhaps not surprisingly, Ackerman's book is now given away free by Creationism.org (69).

False Witnessing and Misrepresentation.

If you want to waste a half hour or so, it's worth scanning though the articles in Creation Ministries website, creationministries.com and totting up the blatant lies and misrepresentations of science and scientists. An example is the claim that that a crude carving of a rhinoceros in front of some leaves on a temple wall at Akor Wat, is really a carving of a stegosaurus (70), an extinct dinosaur with which it bears no similarity.

On example is an article by Jonathan O'Brien, B. Creative Arts, B. App. Sc., a 'former geologist' who gave up a science for which he had no qualifications to write articles for creationists on subjects for which he has no qualifications and no history of research or peer-reviewed publications.

Like other creationists writers, he relies on the fact that his audience will almost certainly never read a scientific paper and even if they did, they wouldn't understand it, so he can be confident his claims will never be fact-checked, so long as they confirm the prejudices of his credulous readers.

Amongst O'Brien's more notorious false claims is the claim that because the cone of a volcano in Mexico grew rapidly, this proves mountains are young (71), ignoring the fact that almost all mountains and mountain ranges are the result of uplift due to plate tectonics, not volcanic activity, as even a 'former geologist' would have known.

But perhaps his most blatant lie is that geologists dated a rock formation to 212 million years old, but when they realised the same rocks had fossilised bird footprints in them, they 'immediately' changed the date to 38 million years to conform to the 'Darwinian narrative' of the evolution of birds

The two papers he refers to are 13 years apart!

The first concerns the alleged dating of wood supposedly found in a rock stratum in Argentina, dated to 212 million years old, but which appeared to have the tracks of wading birds in the same strata – which had not evolved 212 million years ago. The scientists published this believing they had found evidence of birds exiting much earlier than the fossil record suggested.

In his article (72), which he illustrated with a rather nice photo, plagiarised from Shutterstock, without acknowledging their copyright, of a lesser green sandpiper, - a species unknown in Argentina, what O'Brien fails to mention is that closer examination and geomagnetic evidence showed that the stratum with the footprints was not the same formation as that with the wood.

In the best scientific traditions, an anomaly had been identified and was duly investigated, not immediately but ten years later by a team which included the authors of the original paper, which was then withdrawn on 7 August 2013 (73) The scientists had discovered they had made a mistake and promptly informed the scientific community of it. Note how 11 years has become 'immediate' in O'Brien's disinformation polemic.

In the abstract to their first paper in 2002 (74) the scientists said:

> Abstract
> The study of fossilized footprints and tracks of dinosaurs and other vertebrates has provided insight into the origin, evolution and extinction of several major groups and their behaviour; it has also been an important complement to their body fossil record. The known history of birds starts in the Late Jurassic epoch (around 150 Myr ago) with the record of Archaeopteryx, whereas the coelurosaurian ancestors of the birds date back to the Early Jurassic. The hind limbs of Late Triassic epoch theropods lack osteological evidence for an avian reversed hallux and also display other functional differences from birds[7]. Previous references to suggested Late Triassic to Early Jurassic bird-like footprints have been reinterpreted as produced by non-avian dinosaurs having a high angle between digits II and IV and in all cases their avian affinities have been challenged. Here we describe well-preserved and abundant footprints with clearly avian characters from a Late Triassic redbed sequence of Argentina, at least 55 Myr before the first known skeletal record of birds. These footprints document the activities, in an environment interpreted as small ponds associated with ephemeral rivers, of an unknown group of Late Triassic theropods having some avian characters.

In the main text of their paper the scientists described how they arrived at the 212-million-year date:

> The unit at the studied locality has produced remains of Rhexoxylon, a wood morphogenus only reported from Middle to Late Triassic rocks of Gondwana. Concurrently, an interbedded basalt flow located about 80 m

above the track-bearing horizons yielded an $^{40}Ar/^{39}Ar$ plateau age of 212.5 ± 7.0 Myr ago (step-heating analysis on albite crystal) which suggests a Norian–Rhaetian age for the basalt according to recent calibrations of the Triassic–Jurassic timescale.

Note that the presence of the fossilised wood of a type only previously found in Late Triassic formations was only a secondary consideration and not the basis for the dating that O'Brien implies. Indeed, the basis for dating gave a good estimate of the age of the rock tested; the problem was that this was not the rock in which the tracks were found.

As was shown in a subsequent paper the geology of that locality was far more complex than they had realised. This reassessment was not carried out immediately, as O'Brien claims; but was carried out ten years later in 2012 and published in 2013. Nor was it done by a different group of scientists as O'Brien states, but by a team that included the lead author of the original paper, Ricardo N. Melchor.

And the scientists explain how the mistake was made with (75):

> Abstract
> The red bed succession cropping out in the Quebrada Santo Domingo in northwestern Argentina had been for long considered as Upper Triassic–Lower Jurassic in age based on weak radiometric and palaeontological evidence. Preliminary paleomagnetic data confirmed the age and opened questions about the nature of fossil footprints with avian features discovered in the section. Recently the stratigraphic scheme was reviewed with the identification of previously unrecognized discontinuities, and a radiometric dating obtained in a tuff, indicated an Eocene age for the Laguna Brava Formation and the fossil bird footprints, much younger than the previously assigned. We present a detailed paleomagnetic study interpreted within a regional

tectonic and stratigraphic framework, looking for an explanation for the misinterpretation of the preliminary paleomagnetic data. The characteristic remanent magnetizations pass a tilt test and a reversal test. The main magnetic carrier is interpreted to be low Ti titanomagnetites and to a lesser extent hematite.

The characteristic remanent magnetization would be essentially detrital. The obtained paleomagnetic pole (PP) for the Laguna Brava Formation has the following geographic coordinates and statistical parameters: $N=29$, Lon.$=184.5°E$, Lat.$=75.0°S$, $A95=5.6°$ and $K=23.7$. When this PP is compared with another one with similar age obtained in an undeformed area, a declination anomaly is recognized. This anomaly can be interpreted as Laguna Brava Formation belonging to a structural block that rotated about 16° clockwise along a vertical axis after about 34 Ma. This block rotation is consistent with the regional tectonic framework, and would have caused the fortuitous coincidence of the PP with Early Jurassic poles. According to the interpreted magnetostratigraphic correlation, the Laguna Brava Formation would have been deposited during the Late Eocene with a mean sedimentation rate of about 1.4 cm per thousand years, probably in relation to the onset of the Andean deformation.

Had the geologists really been trying to produce evidence to support some assumed Darwinian narrative they wanted people to falsely believe, as O'Brien implies, why would they have published their 2002 paper in which they claimed to have found fossil avian footprints that predated the earliest avian fossils by a considerable margin, only to have to withdraw it later?

Obviously, O'Brien was confident that his claims would simply be regurgitated as proven facts by people for whom fact checking is taking too big a risk of having to change their mind. And how he thinks whether the bird footprints were 38 million or 212 million years old establishes that Earth is 6-10,000 years old is anyone's guess. He probably realises that joined-up thinking is not a strong point of most creationists.

Wouldn't it be refreshing if the creation industry employed people with the same level of integrity as scientists who are prepared to revisit, revise, and retract when mistakes are made? Can you imagine someone who deliberately misrepresents science for a living because he lacks the understanding and/or ability to earn his living as a scientist standing up and admitting he was wrong?

And wouldn't it be refreshing if creationists cult leaders like O'Brien tried to educate their ignorant followers instead of seeing their ignorance as an asset to be carefully nurtured and exploited for all it's worth.

But an outbreak of intellectual integrity and moral probity would spell the end for creationism.

The 'Progression of Man' Fallacy

On of the more insidious and blatant lies promulgated by creationists such as Kent Hovind, or to give him his correct academic title, Mr Kent Hovind, is the familiar illustration of what creationist frauds claim evolutionary paleoanthropologists believe represent the evolution of modern humans from 'Lucy' to modern man.

It depicts, in sequence, with a derisory and inaccurate description of each, 'Lucy', 'Heidelberg Man', 'Nebraska Man', 'Piltdown Man', 'Peking Man', 'Neanderthal Man', 'New Guinea Man', 'Cro Magnon Man' and 'Modern Man'.

The original came from a creationist propaganda comic intended to fool children, called 'Big Daddy' by an infamous liar, 'Jack Chick', in which 'Evolutionists' were always depicted as angry, dirty, scruffy-looking, dishonest, God-hating despicable Atheists with low intellect and no redeeming features and Christians were always clean-shaven, neatly coiffured, smart and sensible.

The cartoon, of course, bears no resemblance to the current thinking regarding the series of intermediate hominins between *Australopithecus afarensis* ('Lucy') from H. habilis, through H. erectus to H. sapiens, possibly via *H. africanus* and *H. heidelbergensis*, with side branches to *H. neanderthalensis* and the Denisovans that the current consensus holds to, so the entire depicted sequence is a lie.

I will deal with each in sequence, together with the description in the drawing:

'Lucy',

The lie:

> All experts agree that Lucy was a 3-foot-tall chimpanzee'. The bigger lie here is that palaeoanthropologists think 'Lucy' was the immediate descendant of a chimpanzee – the so-called 'missing link'.

The truth

> Lucy' is the name given to a particular fossil of *Australopithecus afarensis* (76), of which fossils of some 300 individuals have been found. I have never seen a claim by any palaeoanthropologist that, *Au. afarensis* was a chimpanzee. Nor is it certain the *Au afarensis* was directly ancestral to modern humans. It is a candidate species, but so are others such as *Au. sediba* (77).

Heidelberg Man

The lie:

Built from a jawbone that was conceded by many as 'quite human'.

The truth:

The position of *H. heidelbergensis* is still uncertain but it is normally placed between *H. erectus* and both *H. sapiens* and *H. neanderthalensis*. It is known from specimens found at Sima Los Heusos, Spain, Jinniushan, China, and Kabwe, Zambia. It's true to say that *H. heidelbergensis* is part of the 'muddle in the middle of the Middle Pleistocene, probably as a result of *H. erectus* splitting into a number of species or regional varieties.

Nebraska Man.

With Nebraska Man, we enter the world of Creationist fantasy.

The Lie:

Scientifically built up from one tooth, later found to be the tooth of an extinct pig. (The bigger lie being that 'Nebraska Man' was ever recognised as even a hominin, let alone regarded as ancestral to modern human beings.)

The truth:

The following is based on a blog post I wrote in 2012 (78):

In 1917, Harold Cook, a rancher and geologist, found a fossil tooth which looked vaguely hominid.

In 1922, Henry Fairfield Osborn prepared a paper on it for the journal *Science* in which he named the putative owner of the tooth *Hesperopithecus haroldcookii*. (Ape of the western world - note, ape, not man).

How Creationists Mislead Their Supporters

Following its publication, scientist Grafton Elliott Smith, authored an article for a popular magazine, The popular magazine, *The Illustrated London News*, (not a science journal). This article was illustrated by Amédée Forestier who, with no scientific justification and despite Osborn's protests, based her illustration on 'Java Man' (*Pithecanthropus* now renamed *Homo erectus*).

Osborn declared this illustration "a figment of the imagination of no scientific value, and undoubtedly inaccurate". No scientist had ever coined the term 'Nebraska man' and Osborn had always been careful to avoid making any claim that *Hesperopithecus* was anything more than an advanced primate of some kind.

> "I have not stated that *Hesperopithecus* was either an Ape-man or in the direct line of human ancestry, because I consider it quite possible that we may discover anthropoid apes (*Simiidae*) with teeth closely imitating those of man (*Hominidae*), ...
>
> Until we secure more of the dentition, or parts of the skull or of the skeleton, we cannot be certain whether *Hesperopithecus* .is a member of the *Simiidae* or of the *Hominidae*. (Osborn 1922)

Just as with the Piltdown forgery (see below), jingoistic nationalism became entangled with the science in the popular imagination and America was declared by some to be the place God has chosen to create humans.

Few people, if any, outside America took the find seriously and the scientific world was never more than highly sceptical of the claim. In 1924, George MacCurdy published the two-volume book 'Human Origins' in which he gave *Hesperopithecus*. a mere (and inaccurate) footnote mention with:

> In 1920 [sic], Osborn described two [sic] molars from the Pliocene of Nebraska; he attributed these to an anthropoid primate to which he has given the name H*esperopithecus*. The teeth are not well preserved, so that the validity of Osborn's determination has not yet been generally accepted.

Eventually, further field work was undertaken at the site of the original find in 1925 and further remains were found which confirmed the scientific scepticism. The tooth was found to be that of an extinct peccary, *Prosthenops*, of which other remains were found.

Osborn then retracted his paper in 1927, less than five years after it was published and *Hesperopithecus haroldcookii.* was consigned to the rubbish heap of science, along with so many other briefly considered then discarded ideas. As is usual with science, a hypothesis had been proposed, the facts were considered, further work was carried out, and the hypothesis was falsified and discarded., in a classic vindication of the scientific method.

And that would have been that had it not been for Christian evangelical preachers like Hank Hanegraff and Grant Jeffrey, who have seized on 'Nebraska man' to dupe the credulous world of creationism by claiming that it was a failed attempt to fool people into believing in evolution by dishonest scientists, or at least evidence of how science, especially evolutionary science, is full of mistakes and so should be distrusted.

In fact, 'Nebraska man' illustrates very neatly how science proposes provisional hypotheses, checks the facts, and carries out further research if necessary, and then either falsifies or fails to falsify the provisional hypothesis.

It also shows how the popular press can take a scientific idea and distort it in the popular imagination, often for profit-motives rather than from a desire to inform and educate.

'Nebraska man' also neatly illustrates how creationists seek to mislead their credulous public into misunderstanding any science which would undermine their income and how they will deliberately confuse articles from the popular press with genuine science and will present popular misconceptions as established science.

'Nebraska man' was not, and has never been, a problem for science. 'Nebraska man' is a hoax perpetrated on a gullible creationist public, not by science, but by those who make their living fooling those who are keen and eager to be so fooled. Such is the dishonesty of those who make a living from religious fundamentalism and those off whom these parasites live.

Piltdown Man

The lie:

> The jawbone turned out to be from a modern Ape. (The bigger lie is of course that a forgery which was exposed as such by evolutionary palaeoanthropologists in 1953 is still regarded as part of the evolutionary tree of modern humans)

The Truth:

> The Piltdown hoax is of special interest to me as one of the team that exposed it as a hoax was a personal friend of mine, the late Professor Sir Wilfred LeGros Clarke, President of the Royal Academy and the leading evolutionary palaeoanthropologist of his day. I met him when, as a boy, I delivered his Sunday Newspapers. He

knew I was interested in all aspects of biology and encouraged me by giving me access to hist vast library. He also helped me with a school project on – the evolution of man about which he was the leading authority.

His first wife and my mother were friends.

The following is based in part on a blogpost I wrote in 2012 (79):

Rather than being the embarrassment for evolutionary science that creationists like to pretend, the Piltdown Man hoax was actually a triumph for science. Its exposure was an example of how the Theory of Evolution made an accurate prediction which was confirmed by evidence.

In 1912, February Charles Dawson, an amateur archaeologist contacted Arthur Smith Woodward, Keeper of Geology at the Natural History Museum, stating he had found a section of a human-like skull in Pleistocene gravel beds near Piltdown, East Sussex, England.

That summer, Dawson and Smith Woodward claimed to have discovered a jawbone, some more skull fragments, a set of teeth and some primitive stone tools in the same location, which they connected to the same individual. (80)

Smith Woodward reconstructed the skull fragments and concluded that they belonged to a human ancestor from 500,000 years ago. The discovery was announced at a Geological Society meeting and was given the scientific name *Eoanthropus dawsoni* ("Dawson's dawn-man") (81).

An extensive review of the evidence in 2016 concluded that the forgery was the work of Charles Dawson. (82)

Piltdown's initial acceptance was more to do with jingoistic patriotism than with science, but it needs to be seen in the context of the then fairly recent history of evolutionary theory, the way it had been received by the Anglican Church and how it fitted into English political life.

Contrary to what one would expect, the history of Bible literalism is a fairly recent American phenomenon; in the late nineteenth and early twentieth centuries, the Bible was accepted by most Anglican theologians as allegorical. Just as Jesus had used stories or parables to illustrate his points, so too had the early prophets. The creation story in particular was allegorical and few Anglican theologians seriously doubted that evolution was the way God had really created humans. The Anglican Church had, since the latter half of the 19th century, been moving away from the belief that the Bible was literal truth in view of the accumulating scientific evidence that it was not.

Another fallacy, and fundamental misunderstanding of Darwinian evolution, was the belief that there was an aim to evolution. Clearly, if God had created evolution, He had done so for a purpose. Obviously, His intention was to put European humans, led by Englishmen, at the apex, with other races occupying inferior positions lower down the 'ladder of evolution'. This was self-evidently so because this had been the result. How could it be otherwise?

The authorities of the time of course encouraged this fundamental error, with the enthusiastic endorsement of the Church of England, because it justified the Empire and placed Englishmen at the pinnacle. Just what the ruling classes required of science. What could be more convincing than a solid scientific theory underpinning the whole edifice?

And this church-endorsed misconception of evolution in general and of human evolution in particular was to have catastrophic consequences when taken up with fanatical zeal by Catholic zealots in Germany, Portugal, Spain and Italy and, it has to be said, certain sections of the ruling classes in other European and New World countries, including England and the USA. It gave rise to fascism as an expression of European racial and Christian cultural supremacy.

So, the notion that God had chosen rural southern England as the place in which to create humans and evolve them into their present state of perfection, seemed perfectly natural and self-evident, and the vindication of everything a true Englishman held most dear. Humans were God's special creation; evolution was the way he had done it and of course, England was the place He had chosen to do it. God was in his Heaven, and all was well with the world. Rule Britannia!

But gradually, as discoveries were made from other parts of the world - from Asia and Africa in particular - and these were fitted into the emerging picture of human evolution, not directly from a monkey, as Piltdown seemed to show, but from a common ancestor with the other great apes, so Piltdown became more and more anomalous. Far from 'proving' human evolution from monkeys in southern England, Piltdown was out of place; a nuisance to be explained away, and, as some bold palaeoanthropologists like Sir Wilfrid Le Gros Clarke, Joseph Weiner and Kenneth Oakley thought, an outright forgery. (83)

How Creationists Mislead Their Supporters

Professor Sir Wilfrid Le Gros Clark told me that he and his team in Oxford were sure that Piltdown MUST be a forgery. It was simply too far out of place to be genuine. No amount of stretching the known facts could accommodate it. The prediction that Piltdown MUST be a forgery was thus a falsifiable prediction of evolutionary theory.

Sir Wilfrid told me that as they started to drill into the skull, using a small dental drill, they saw a wisp of smoke and smelled burning bone and knew then, within a few seconds of beginning their examination, that they had a forgery and that their prediction had been vindicated. Subsequent microscopic examination revealed the marks of the forger all over it, including the teeth with file marks.

So, the debunking of Piltdown Man by evolutionary palaeontologists, far from being an embarrassment to science, was a triumph for it and for the theory of evolution. The Theory of Evolution had made a falsifiable prediction and had been entirely vindicated. Science constantly re-examines and re-assess its theories; discovers its mistakes and moves on to greater things. Even an elaborate hoax having the support of the church and political establishment will eventually be exposed.

And the debunking of Piltdown also dealt a blow to the church-endorsed, and religiously inspired notion of human supremacy and of European racial superiority. The obscenity of social 'Darwinism' is of course a religiously inspired forgery of which the Anglican and Catholic churches in particular and Christianity in general should be thoroughly ashamed and embarrassed. No wonder they like now to pretend that Piltdown was, and continues to be, an embarrassment for science.

When will we see them reviewing their dogmas, reassessing them and correcting their mistakes? When will we see the Anglican and Catholic churches apologise for their misrepresentation of Darwinian evolutionary theory and the obscene dogma of white racial supremacy to which it led, and which is still promulgated in some parts of the world, particularly in the USA where Christian white supremacy is again rearing its ugly head and enthusiastically embracing social 'Darwinism'?

Peking Man

The lie:

> Supposedly 500,000 years old but all evidence has disappeared. (The same bigger lie is of course that Piking Man was ancestral to modern *H. sapiens.*

The truth:

> Peking man' was a subspecies of *Homo erectus* of which 15 partial craniums, 11 lower jaws, many teeth and some skeletal fragments were recovered from a site at Zhoukoudian near Beijing, China between 1929 and 1937.
>
> Several casts of the originals remain and some teeth at Upsala University, Sweden, however, the original fossils themselves went missing in 1941 somewhere in Northern China which was then under Japanese occupation, whilst en route to the USA. A substantial reward for their recovery has been offered by the Chinese Government. In 1966 however, more fragments were found at the same site.
>
> Quite where *Homo erectus pekinensis* fits into the evolution of modern humans is still uncertain. Some authorities suggest it may be ancestral, at least partially,

to modern Chinese, maybe with interbreeding with early *Homo sapiens*, however genetic studies show that modern Chinese fall within the range of diversity of all modern humans, suggesting there was little if any interbreeding between modern humans and *Homo erectus*.

No one has ever suggested that *Homo e. pekinensis* somehow fits in between 'Piltdown Man' and Neanderthals in the scientific account of human evolution as the drawing shows. Without the specimens for modern dating techniques all we have is estimates ranging from 230,000 to 780,000 years – which is consistent with a local variety of *H. erectus* but how *H. e. pekinensis* fits in with the Denisovans is still unclear.

Neanderthal man

The lie:

> At the Int'l Congress of Anthropology (1958) Dr. A.J.E. Cave said his examination showed that this famous skeleton found in France is that of an old man who died of arthritis. The clear implication there is that there is only one skeleton of Neanderthal and that the opinion of one man in 1958 is the definitive word on the subject. The original Neanderthal was discovered in the Neander Valley (Neander Thalle) in Germany

The truth:

> In fact, there are dozens of fossil remains of Neanderthals, from several locations in Europe and the Middle East, and we now have a fully sequenced genomes not of one but of several. There is little doubt that they were a now-extinct species of hominin that evolved from *H. erectus*, possibly via *H. heidelbergensis* or *H. antecessor* in Eurasia and were the dominant hominins in Europe for some 250,000 years

with a sophisticated hunter-gatherer culture and possibly religious ceremonies. However, few if any palaeoanthropologists think they were the direct ancestors of *H. sapiens*, with whom they co-existed and interbred in Eurasia.

New Guinea Man

The lie:

> Dates way back to 1970. This species has been found in the region just north of Australia.

The truth:

> This appears to be a figment of 'Jack Chick's fertile imagination as there is no recognisable species of hominin described as 'New Guinea Man'', and even if there were, no sane paleoanthropologist would place it between Neanderthals and modern humans. The people of Papua New Guinea are *H. sapiens*, and like several other people from Southeast Asia, Austronesia and Oceania, have about 3% Denisovan DNA, showing interbreeding between the two contemporaneous species of hominin.

Cro Magnon Man

The lie:

> One of the earliest and best-established fossils is at best equal in physique and brain capacity to modern man... so what's the difference?

The truth:

> Co magnon was a culture, not a species. The people who lived at the Abris du Cro Magnon (meaning, 'large rock overhang' in Occitan – the language then spoken in that part of Southwest France). No-one has ever claimed

they were distinct species to modern humans or even ancestral to them, and they were certainly not ancestral to the modern humans of Africa, Asia or the Americas. They were an early European cultural group, the name for whom is sometimes misapplied to other early modern humans. The question, '... so what's difference?' simply betrays the ignorance of the questioner.

Modern Man

The lie:

> This genius thinks we came from a monkey!
>
> The more subtle lie in that drawing is that 'modern man' is clearly depicted as a European, pandering to the racist notion of European supremacy as the end point of human evolution, or at least that's what scientists think.

The truth:

> Modern humans are a species of ape that split from the chimpanzee line in Africa some 6 million years ago. There is abundant fossil evidence recording our evolution, and our DNA confirms our relationship to the other African apes and further back to the catarrhine simians that evolved into the anthropoid apes.
>
> Ironically, 'Jack Chick' is embracing social 'Darwinism' probably because of an underlying assumption of racial superiority, prevalent amongst White American evangelical Christians.
>
> No amount of ridiculing childish parodies, throwing stones at carefully constructed straw men and misrepresentations of the scientific evidence, is going to change that.

Quote Mining

Quote mining is a common tactic of creationists intended to give the impression that an authority figure had expressed support for creationism or against the TOE, or had somehow inadvertently admitted he didn't really believe in evolution.

It is a form of false witnessing that betrays a motive other than evangelising for a religious faith – which forbids bearing false witness. It depends for its success as a tactic on the knowledge that creationist will never fact check in case they discover they are wrong.

One form of quote mining is to present an opinion made in the past as though it were current thinking, despite the fact that science has progressed since the comment was made. Another is to quote a creationist but to claim the author was an evolutionary biologist of other expert in a field of science.

A common deception is to take the opening paragraph of an article or scientific paper where the problem addressed is stated, and present that at the opinion of the authors, ignoring the fact that the rest of the article gives the solution to the problem. Darwin's writing is full of instances where he outlines a problem for what was then conventional biology, then spends the rest of the chapter explaining how his theory solves the problem, because that was his style of writing and his problem-solving approach to science.

Some of these are taken from The Talk Origins Archive. (84) They can be found scattered throughout the creationist literature and websites.

Lack of Identifiable Phylogeny

The quote mine :

> "It is, however, very difficult to establish the precise lines of descent, termed phylogenies, for most

organisms." (*Evolving: The Theory and Process of Organic Evolution*, 1978, p. 230) (85)

The truth:

In full context the authors actually said:

"It is, however, very difficult to establish the precise lines of descent, termed phylogenies, for most organisms. A direct method of tracing phylogenies has been to trace a series of fossils that resemble each other but show a sequence of changes leading through time from an ancestral to a descendant form. Relationships among the fossils are thus judged by their relative ages and their morphological resemblances and differences. This works well when abundant fossils are available in a continuous record, but unfortunately the fossil record is quite incomplete. Most animals have no easily fossilizable hard parts, and only a small fraction of animals with shells or bones are actually preserved as fossils. For most lineages we have to employ more indirect methods of phylogenetic reconstruction."

The quote mine;

"Undeniably, the fossil record has provided disappointingly few gradual series. The origins of many groups are still not documented at all." (Futuyama, D., *Science on Trial: The Case for Evolution*, 1983, p. 190-191) (86)

The truth:

"Contrary to Creationist claims, the transitions among vertebrate species are almost all documented to a greater or lesser extent. Archeopteryx is an exquisite link between reptiles and birds; the therapsids provide an abundance of evidence for the transition from reptiles to mammals. Moreover, there are exquisite fossil links between the crossopterygian fishes and the amphibians (the icthyostegids). Of course, many other ancestor-descendent series also exist in the fossil record. I have mentioned (Chapter 4) the bactritid-ammonoid transition, the derivation of several mammalian orders from condylarthlike mammals, the evolution of horses, and of course the hominids.

"Undeniably, the fossil record has provided disappointingly few gradual series. The origins of many groups are still not documented at all. But in view of the rapid pace evolution can take, and the extreme incompleteness of fossil deposits, we are fortunate to have as many transitions as we do. The creationist argument that if evolution were true we should have an abundance of intermediate fossils is built by denying the richness of palaeontological collections, by denying the transitional series that exist, and by distorting, or misunderstanding, the genetical theory of evolution.".

The quote mine;

"The main problem with such phyletic gradualism is that the fossil record provides so little evidence for it. Very rarely can we trace the gradual transformation of one entire species into another through a finely graded sequence of intermediary forms." (Gould, S.J. Luria, S.E. & Singer, S., *A View of Life*, 1981, p. 641) (87)

The truth:

Later, on the same page:

There is an alternative, however. Perhaps the fossil record is not so hopeless, and the observation of no change within species and sudden replacement between them reflects evolution as it actually occurs. Recall Chapter 26: Large, successful, central populations are resistant to evolutionary change. Small, isolated,

How Creationists Mislead Their Supporters

> marginal populations may speciate. The process of speciation, though slow to a human observer (hundreds or thousands of years), is geologically fleeting. In most geological situations, and at most rates of sedimentation, a thousand years translates into a single bedding plane, not a thick sequence of rock. Thus, if speciation is the dominate mode of evolution, we should expect to see exactly what we do see: the unchanging species represents a successful central population; its sudden replacement by a descendent records the migration into the ancestral area of a descendant that arose rapidly in a small population at the edge of the ancestor's geographical range. Thus, it is possible that most evolution occurs in the mode of speciation and that phyletic evolution is relatively unimportant.

The quote mine: (apparently copied and posted from another quote mined source without realising it was from an anti-creationist book from a section ironically pointing out how creationist quote mine). It is also a misquote since the initial word 'It' should not have been capitalised:

> "It should come as no surprise that it would be extremely difficult to find a specific fossil species that is both intermediate in morphology between two other taxa and is also in the appropriate stratigraphic position." (Cracraft, J., "Systematics, Comparative Biology, and the Case Against Creationism," 1983, p. 180) (88)

The truth:

> In using selected quotations of palaeontologists to buttress their own position, creationists have unwittingly entered one the most controversy theoretical and methodological debates in contemporary palaeontological systematics. Inasmuch as this debate in has ensued for over a decade in the scientific literature, it is surprising that the creationists have not mentioned

its existence (either the creationists are unfamiliar with the scientific literature or they have failed to understand the importance of that literature or they have simply chosen to ignore the problem and adopt a strategy that promotes their theological, not scientific position). The debate centres on the scientific methods used to postulate and test hypotheses of ancestral-descendant relationship. Traditionally, palaeontologists, including most quoted by the creationists, have had a conviction that the stratigraphic position of the fossil taxa is a primary criterion with which to postulate ancestral-descendant relationships, whereas recent critics of this methodology have stressed the importance of a critical analysis of morphological characteristics ([deleted references]). If the stratigraphic position of a fossil [p. 179 | p. 180] is an important criterion for recognizing it as an ancestor, it should come as no surprise that it would be extremely difficult to find a specific fossil species that is both intermediate in morphology between two other taxa and is also in the appropriate stratigraphic position. There is no doubt the reason for many of the quotes cited by the creationists about the prevalence of gaps, but other citations or distortions, tailored to suit the creationists' own purposes. For example, in 1972 Schaeffer, Hecht, and Eldredge (89) published an influential paper in which they were critical of palaeontological methodology about the construction of ancestral-descendant hypotheses. In support of his argument that there are no transitional forms, Gish (1979, p. 169) (90) quoted from a review of that paper.

Karl Popper

Karl Popper, the philosopher, who basically said, you can't prove a negative —something that creationists repeatedly demand science does - is quoted here (91) by creationist, Russel Kranz, as saying,

"Darwinism is not a testable scientific theory, but a metaphysical research programme."

However, Popper also had this to say about the TOE:

"And yet, the theory is invaluable. I do not see how, without it, our knowledge could have grown as it has done since Darwin. In trying to explain experiments with bacteria which become adapted to, say, penicillin, it is quite clear that we are greatly helped by the theory of natural selection. Although it is metaphysical, it sheds much light upon very concrete and very practical research. It allows us to study adaptation to a new environment (such as a penicillin-infested environment) in a rational way: it suggests the existence of a mechanism of adaptation, and it allows us even to study in detail the mechanism at work. And it is the only theory so far which does all that."

And:

"When speaking here of Darwinism, I shall speak always of today's theory – that is Darwin's own theory of natural selection supported by the Mendelian theory of heredity, by the theory of the mutation and recombination of genes in a gene pool, and by the decoded genetic code. This is an immensely impressive and powerful theory. The claim that it completely explains evolution is of course a bold claim, and very far from being established. All scientific theories are conjectures, even those that have successfully passed many severe and varied tests. The Mendelian underpinning of modern Darwinism has been well tested, and so has the theory of evolution which says that all terrestrial life has evolved from a few primitive unicellular organisms, possibly even from one single organism.

And:

"I still believe that natural selection works in this way as a research programme. Nevertheless, I have changed my mind about the testability and the logical status of the theory of natural selection; and I am glad to have an opportunity to make a recantation. My recantation may, I hope, contribute a little to the understanding of the status of natural selection.

Not exactly the ringing endorsement of creationism that out little false witnessing quote miner wanted us to believe.

And of course, Duane Gish, who never misses an opportunity to mislead his followers cited Popper and claimed he meant evolution is religious mysticism and therefore false (unlike the mystical religion, Christianity) and is being increasing recognised as such more and more.

Duane Gish ceased his galloping in 2013, since when belief in creationism has slumped even further in the American opinion polls.

Charles Darwin

Needless to say, creationists would love nothing more to have the top of their demonology, Charles Darwin, telling 'evolutionists', who hang on every word that he wrote as reveal truth by their prophet, that they are wrong, If he 'accidentally' let slip that he didn't believe a word of it himself, so much the better, so wilful misquotes or out of context passages by him are gleefully seized upon, and dutifully copied by other creationists.

One such internet creationists is Tracy Zdelar who either through ignorance or mendacity has posted the following quote mines on her website, Hall of Fame Moms (92). Tellingly, although she mentions an update to her post that Darwin quotes are taken out of context, she fails to correct or retract them but sticks to her false witnessing:

Quote mine:

> "If it could be demonstrated that any complex organ existed which could not possibly have been formed by numerous, successive slight modifications my theory would absolutely break down."

How Creationists Mislead Their Supporters

What Darwin really said (93)

> "If it could be demonstrated that any complex organ existed which could not possibly have been formed by numerous, successive slight modifications my theory would absolutely break down. *But I can find out no such case.*"

This quote mine was also cited by Casey Luskin in a book, *Discovering Intelligent Design: A Journey into the Scientific Evidence* (94), advocating for intelligent design, presumably because he felt his argument was too weak so needed the apparent support of none other than Charles Darwin.

Quote mine:

"Such simple instincts as bees making a beehive could be sufficient to overthrow my whole theory."

What Darwin really said:

> "The subject of instinct might have been worked into the previous chapters; but I have thought that it would be more convenient to treat the subject separately, especially as so wonderful an instinct as that of the hive-bee making its cells will probably have occurred to many readers, as a difficulty sufficient to overthrow my whole theory. I must premise, that I have nothing to do with the origin of the primary mental powers, any more than I have with that of life itself. We are concerned only with the diversities of instinct and of the other mental qualities of animals within the same class."

Darwin is saying here that, although some people might think hive-making bees are a problem for his theory, he does not agree with them.

Not only taken out of context but altered to suit the purpose of deception. False witnessing at its worst- lying about what was actually said.

Note also that Darwin is saying here that his theory has nothing to do with the origin of life, but with how diversification happens.

Quote mine:

> Often a cold shudder has run through me, and I have asked myself whether I may have not devoted myself to a phantasy.

What Darwin really said:

The context of this quote is a letter to his friend and mentor Charles Lyell when his 'Origins; was in the process of being published and thanking him for his support.

"I rejoice profoundly that you intend admitting doctrine of modification in your new Edition. Nothing, I am convinced, could be more important for its success. I honour you most sincerely: — to have maintained, in the position of a master, one side of a question for 30 years & then deliberately give it up, is a fact, to which I much doubt whether the records of science offer a parallel. For myself, also, I rejoice profoundly; for thinking of the many cases of men pursuing an illusion for years, often & often a cold shudder has run through me & I have asked myself whether I may not have devoted my life to a phantasy. Now I look at it as morally impossible that investigators of truth like you & Hooker can be wholly wrong; & therefore I feel that I may rest in peace."

Rather than saying what the quote mine implies by omitting the following sentences, Darwin is actually saying exactly the opposite.

How Creationists Mislead Their Supporters

The following quote is repeated ad nauseum despite even Ken Ham's Answers in Genesis, grudgingly accepting that it's taken out of context (95). For example, here (96)

This one will be regurgitated regularly in any Science vs. Creationism group in the online social media.

The Quote mine:

> "To suppose that the eye, with all its inimitable contrivances for adjusting the focus to different distances, for admitting different amounts of light, and for the correction of spherical and chromatic aberration, could have been formed by natural selection seems, I freely confess, absurd in the highest possible degree."

What Darwin actually said:

> "To suppose that the eye with all its inimitable contrivances for adjusting the focus to different distances, for admitting different amounts of light, and for the correction of spherical and chromatic aberration, could have been formed by natural selection, seems, I freely confess, absurd in the highest degree. When it was first said that the sun stood still and the world turned round, the common sense of mankind declared the doctrine false; but the old saying of Vox populi, vox Dei, as every philosopher knows, cannot be trusted in science. Reason tells me, that if numerous gradations from a simple and imperfect eye to one complex and perfect can be shown to exist, each grade being useful to its possessor, as is certainly the case; if further, the eye ever varies and the variations be inherited, as is likewise certainly the case; and if such variations should be useful to any animal under changing conditions of life, then the difficulty of believing that a perfect and complex eye could be formed by natural selection, though insuperable by our imagination, should not be considered as subversive of the theory. How a nerve

> comes to be sensitive to light, hardly concerns us more than how life itself originated; but I may remark that, as some of the lowest organisms, in which nerves cannot be detected, are capable of perceiving light, it does not seem impossible that certain sensitive elements in their sarcode should become aggregated and developed into nerves, endowed with this special sensibility (97).

Not only is Darwin emphatically not saying what the quote mine implies, he was saying exactly the opposite – that his TOE explains how it came about.

He is also warning against the argument from ignorant incredulity that is a normal part of creationist rhetoric because our intuition, especially when it comes from ignorance, is simply not the best measure of reality and is no match for observation and reason of science.

Deliberate misrepresentation of science

There is nothing more that creationists love more than the thought that there is solid scientific support for their superstition, so creationists site often include references to actual scientific papers in their articles, relying on the fact that creationists rarely fact check by reading the article and if they do, they don't understand what they're reading.

It is a rule, almost amounting to a law of internet debate with creationist, that a given reference will not say what the creationists claims it says and will flatly contradict them. The trick seems to be to do a quick trawl with a few key words of phrases, then cite anything found. After just reading the title. The clear intention there is to create the illusion that the claim being made has scientific support.

However, there are frequent examples of someone posing as an expert, or indeed being an expert in a subject, citing an article that contradicts his claim, of which he should have been aware and probably was,

How Creationists Mislead Their Supporters

A good source of examples of this is Ken Ham's site, Answers in Genesis, especially his *The 10 Best Evidences [sic] from Science That Confirm a Young Earth* (98). It starts off well with the classic circular reasoning – the Bible is God's word because it says in the Bible that it is God's word – and goes downhill from there.

Most of the actual 10 arguments are written by AiG staffer and real-life geologist, Andrew Snelling. Snelling is notorious for producing misleading articles for creationists 'proving' Earth is just a few thousand years old, while putting his name to academic geology textbooks that say the opposite. For example, in an authoritative two volume work entitled "Geology of the Mineral Deposits of Australia and Papua New Guinea" (ed. F E Hughes), published by the Australasian Institute of Mining and Metallurgy, Melbourne, (99) Snelling says:

> "The Archaean basement consists of domes of granitoids and granitic gneisses (the Nanambu Complex), the nearest outcrop being 5 km to the north. Some of the lowermost overlying Proterozoic metasediments were accreted to these domes during amphibolite grade regional metamorphism (5 to 8 kb and 550° to 630° C) at 1870 to 1800 Myr. Multiple isoclinal recumbent folding accompanied metamorphism" (Page 807) [my emphasis].

In the first of Ham's '10 top Evidences [sic]. Snelling claims scientific support for the notion that there is not enough sediment on the ocean floor for Earth to be more than a few thousand years old (despite the fact that as a geologist he will be familiar with plate tectonics and how plates are subducted beneath other plates, taking the ocean floor sediment with them. He will also be aware that the ocean floors only date back to the breakup of Pangea 175 million years ago.

But he cites a paper, by John D. Milliman and James P. N. Syvitski, "Geomorphic/Tectonic Control of Sediment Discharge to the Ocean: The Importance of Small Mountainous Rivers, *"The Journal of Geology* 100 (1992): 525–544 (100) as the source for his statement that:

> Every year water and wind erode about 20 billion tons of dirt and rock debris from the continents and deposit them on the seafloor".

However, when we 'cheat' and check that reference – something that we are not supposed to do, apparently - what we see in the abstract is:

> Before the proliferation of dam construction in the latter half of this century, rivers probably discharged about 20 billion tons of sediment annually to the ocean. Prior to widespread farming and deforestation (beginning 2000-2500 yr ago), however, sediment discharge probably was less than half the present level.

Spot the difference.

The second of the 'Top Evidences [sic] again by Snelling, pulls the same trick in regard to scientific evidence that rocks can only bend when wet (so bent strata must have been done in the biblical flood). Snelling uses the false equivalence fallacy here to compare the solidification of rock to the setting of cement.

As a geologist Snelling will have been aware that the formation of rock bears no resemblance to the setting of cement, which is a rapid chemical process in which CO_2 reacts with quicklime (CaO) via calcium hydroxide ($Ca(OH)_2$) to give calcium carbonate ($CaCO_3$). The formation of rock, however, takes place slowly over time as the sediment accumulates and places the lower strata under pressure which squeezes the water from between the particles, to be replaced by crystalised minerals binding then together. During this process, there is ample time for the forming rocks to be deformed by geological forces.

How Creationists Mislead Their Supporters

To lend credence to his claim, Snelling cites a book by Richard E. Goodman. - Introduction to Rock Mechanics (New York: John Wiley and Sons, 1980) (101),

However, if we again cheat and check we will see that Goodman devotes an entire chapter (6 – Deformability of Rocks) to explaining how and under what circumstances rock bends under stress. The chapter contains lots of complicated mathematics and diagrams, but nowhere does it say rocks can only bend when wet. In fact, it flatly contradicts Snelling's claim.

When you show the world you know you need to lie for your faith, you show the world you know your faith is for fools who will believe falsehoods.

Refuting Creationism

The Fine-Tuned Fallacy

The claim that the Universe is fine-tuned for life is frequently proposed as 'proof' that there must have been a designer, but it is based on a number of false premises, faulty logic, and downright deceptions.

The Anthropic Principle

Ask yourself the simple question: could we be discussing the nature and origins of a Universe in which life could not exist?

Of course not.

The fact that we can discuss it means we exist in a Universe which is suitable for our existence. Does that mean the Universe must have been purposefully designed so that we could exist, or is it possible that a different Universe could have existed (or could exist now according to the multiverse hypothesis)?

This is a form of the false logic in trying to calculate the probability of something that has happened happening. Of course, the probability of a past event having happened is certainty (expressed as 1 in probability theory). It has happened.

A Misuse of statistics

The argument is intelligently designed to give a probability so low that it approaches zero (impossible).

It is a misuse of statistics to try to calculate the probability of a past event the same way you would calculate the probability of a future event.

Take the simple example of dealing hands of cards. Take a standard pack of 52 cards and deal out four hands of 13 cards each. Make a note of the exact order in which the cards were dealt.

Do that 9 more times to give a total of 40, 13-card hands. Now calculate the probability (P) of exactly those 40 hands being dealt in that exact order.

The answer is:

$$P = \frac{1}{\frac{520!}{(13!)^{40}}}$$

For those not familiar with probabilities, the '!' (factorial) character means multiply that number by every other smaller number between it and 1, so $13! = 13 \times 12 \times 11 \ldots \times 2 \times 1$.

Factorial number very quickly become astronomical so 520! is too large to be written in a book of this size. It is a number with about 1,189 digits before the decimal point (1 million has 7 digits and 1 billion has 10). It approaches the number of elementary particles in the known universe.

But it is divided by 13! Raised to the 40th power, in other words multiplied by itself 40 times. Nevertheless, it is still a massively vast number.

So, the formula becomes:

$$\frac{1}{Really\ big\ number} = Really\ small\ number$$

The result is a probability so small that it approaches zero (impossible) and again is so small that it would need several pages of zeros following the decimal point to express it.

Voilà! Using creationist logic, you just 'proved' you cannot deal hands of card because it takes a god who can do the impossible to do it (the locally popular one, naturally.)

The error is in assuming the outcome was exactly the outcome intended. Had you written down the exact order of the cards before dealing them then that would have been impressive, to say the least.

The Fine-Tuned Fallacy

But the fine-tuned argument breaks down in other ways, not just in its deceptive abuse of maths.

The Vanishingly Small Probability Tactic

Creationists take a range of so-called 'parameters' which supposedly determine the nature of the universe, and particularly its ability to make atoms which aggregate into molecules and then into living organisms.

First, let us consider how a scientist would determine the probability of any one thing having a particular value. Let us take the probability of any randomly selected American adult male being 5'10" tall. What they would do is, take a randomly selected sample of adult American males and measure their height. They would then work out what proportion of those men were five'10" tall and that would be the approximate probability of any randomly selected American man being 4'10" tall.

Now, supposing they selected just 10 males in their sample, obviously the chance that they might not be a representative sample of all American adult male, so the confidence in that average would be low. For a sample size of 100 the confidence would be higher but not high enough to base scientific calculations on it. At a sample size of 1000, confidence increases but is still too low, so the sample size, if truly random needs to be something like 10,000.

This is because of fundamental mathematical rules that govern statistics:

The Law of Large Numbers:

This law states that as the sample size increases, the sample mean converges on the population or true mean. When you take a large enough sample, the sample mean is more likely to reflect the population mean accurately because the impact of random errors or outliers diminishes.

The Central Limit Theorem (CLT):

According to the CLT, the distribution of sample means will approach a normal distribution as the sample size increases, regardless of the original distribution shape. For a sufficiently large sample, this normal distribution of sample means has a standard deviation called the *standard error*, which decreases with increasing sample size. The standard error, given by:

$$\frac{\sigma}{\sqrt{n}}$$

(where σ is the population standard deviation and n is the sample size), becomes smaller with a larger sample size, indicating that the sample mean is clustering closer to the population mean.

Reduced Margin of Error:

With larger sample sizes, confidence intervals around the sample mean become narrower because of the reduced standard error. A narrow confidence interval implies greater precision in estimating the true population mean, thereby increasing confidence in its accuracy.

But it's not the mean we're interested in; it's the probability of any sample have a particular value, and we can calculate that from the mean and standard deviation (σ) to calculate the 'Z-score' which is then used to find the probability of that particular value from a 'Z-table' or by a statistical method that is outside the scope of this book. To calculate the 'Z-score' for a particular value (X) statisticians use:

$$Z = \frac{X - \mu}{\sigma}$$

Where μ is the mean and σ is the standard deviation. This means that the Z value gets smaller as the standard deviation gets smaller (in other words, as the sample size gets larger).

The Fine-Tuned Fallacy

The take-away point from that brief lesson in statistics is that you can only begin to calculate the probability of a measure having any particular value from a large sample size.

So, do creationists take a randomly selected sample of Universes and measure the value of their parameters?

Guesses and Gobbledygook.

Of course not. They take a sample size of 1 (this universe) and guess the probability of any parameter having its current value. They have no more idea than fly what that probability is; for all they know it could be 1 (certainty).

There is no way that they can accurately assess the probability of any of the universe's parameters having its current value, based on the measurements from a sample of 1 They simply make it up! They start with the answer they need and work backward.

How, for example did they arrive at the probability of the velocity of light in a vacuum being exactly 299,792,458 metres per second? Did they find examples of it having different values so they could produce a distribution curve of different values, or examine a randomised sample of different universe? Of course not! They simply assumed the probability would be exceedingly small, when in fact it is probably 1 (certainty).

If the probability of those parameters having their current value in this universe is 1 (certainty) then the entire probability of the universe being as it is, is 1 (certainty)

The trick is to pluck a random selection if very small numbers, out of thin air, multiplied them together to give a vanishingly small number and divided 1 by that number and declare that to be the probability of the universe being as it is – so unlikely that the same 'god who can do impossible things' who has to deal hands of cards, must have designed the Universe.

Wrap that all up in some sciencey-sounding gobbledygook and some impressive mathematical formulae, and the audience will go away happy that you've just proved their god exists., although none of them will have understood the mathematical formulae or the statistical techniques well enough to see the flaw in the argument or notice the sleights of hand.

The King has a wonderful new suit of clothes! The 'experts' said so! And who is going to be brave enough to point out that his clothes are an illusion?

And, funniest of all, as we've seen Young Earth Creationists shoot themselves in the foot with this argument because one of their 'explanations' for radiometric dating showing Earth is very much older than they believe, is that radioactive decay rates were much higher in the past than they are now, and radioactive decay rates depend on one of these 'finely-tunes' parameters, the weak nuclear force.

For radioactive decay rates to have been higher in the past, the weak nuclear force would have to have been significantly weaker than it is now, so, by their own 'fine-tuned' argument the Universe would not have been fine-tuned for life when life was being created as atoms and molecules could not have existed in what would have been a soup of elementary particles and electromagnetic radiation in an opaque Universe. 'Let there be Light' would have been meaningless in such a universe.

Why The Bible Cannot Be Taken Seriously.

Although their over-riding preoccupation is with attacking the Theory of Evolution and especially Charles Darwin as its prophet and the 'Gospel of Creationism' 'The Origin of Species', creationists have a couple more obsessions on which their notion of the Bible being a literal science and history textbook rests.

They are fixated on trying to rubbish the geological and cosmological evidence not only of an old Earth in an older Universe, but the evidence that there never was a genocidal flood in which two of each species survived on a wooden boat for about a year, and that there never were enough people to build a tower to above the clouds as in the Tower of Babel myth.

Other absurd Bible stories they need to defend are those of Jonah, Lot and the Cities of the Plain, The Canaanite genocide and, of course being fundamentalist Christians, they need to defend every jot and tittle of the New Testament, even those that contradict each other.

I don't intend to go through the entire Bible and refute every story in it, but there are some in Genesis especially that seem to pre-occupy creationist more so than the others. The most important being the story of a genocidal flood just a few thousand years ago. The dates vary but I have seen claims that it was in about 2000 BCE, in other words about 4,000 years ago.

Noah and The Genocidal Flood

The changing definition of 'evolution'

In the 1980s, it was standard creationists dogma that evolution didn't happen and was actually forbidden by the Second Law of Thermodynamics, (2LOT) although few if any of them could quote the 2LOT or explain how it applied to evolution by natural selection, but that was it; a blanket denial that evolution ever happened.

The it slowly dawned on their cult leaders that the Noah's Ark story needed to explain how two (or seven, depending on the version of the story) of 220 million species (including those that have subsequently gone extinct (some 99% of all species) could have been accommodated on a wooden boat built in a few weeks by an almost 600-year-old [sic] man, and his three octogenarian sons conceived when Noah and his wife were 500 year old, allegedly, with no history of boat-building

This was clearly impossible as a boat that large would be crushed under its own weight and would snap in two if it was met head on by a couple of waves of the right wavelength, to lift the bow and stern but leave the middle unsupported (there are reasons why Medieval galleons were never the size of super-tankers),

So, the prohibition of evolution by the 2LOT was quickly abandoned, and only kept in reserve for the special occasions when they need to refute actual evidence of evolution, and warp speed evolution was invented so the Bible story could be revised to mean only two (or seven) of each 'kind' were saved on the Ark, followed by a rapid period of evolutionary diversification that somehow no-one thought fit to record anywhere, but which involved animals giving birth to new genetically-isolated species several times every generation, with a big enough population to avoid a fatally narrow genetic bottleneck.

And so, those modern creationists who have managed to keep up with developments will happily accept their warp-speed version of 'micro-evolution' while their version of 'macro-evolution is still strictly prohibited by 2LOT.

Recall how this distinction is artificial as I explained in an earlier chapter.

What I will concentrate on now is the absurdity of the flood/Ark myth and the singular lack of any supporting evidence for the tale.

Why The Bible Cannot Be Taken Seriously

A whole lot of Water

Firstly, recall that the Bible says Earth was covered in water to a depth sufficient to cover the highest mountain to a depth of 15 cubits (about 45 feet). Mount Everest is some 29,000 feet high, so that gives us a basis for calculating how much water fell as rain and erupted from the 'Fountains of the Deep, whatever they are.

Was it a boat or a box?

First the dimensions of this alleged wooden boat (and note that the building instructions never relate to a boat as we know it, but always to an 'Ark' which, when we come to the story of the 'Ark of the Covenant', means box. With no power, not even sails, and no destination this box would have needed no fore and aft; no means of steering it, no prow, and no keel, and none is mentioned.

> And this is the fashion which thou shalt make it of: The length of the ark shall be three hundred cubits, the breadth of it fifty cubits, and the height of it thirty cubits. A window shalt thou make to the ark, and in a cubit shalt thou finish it above; and the door of the ark shalt thou set in the side thereof; with lower, second, and third stories shalt thou make it. (Genesis 6: 15-16)

Bear in mind that a cubit was the length of an adult man's forearm, so we are talking about a box which is 900 feet long, 150 feet wide and 90 feet tall. It has three decks but only a single window, 3 feet by 3 feet (i.e. 9 square feet). Clearly two of the decks do not have a window.

Presumably, the door is at ground level as there is no mention of steps or a ramp. This means the door would have been submerged when the Ark was afloat.

But back to that deluge which the Bible tells us lasted 40 days and 40 nights. Simple maths tells us that to reach 29,000 feet (the height of Mount Everest) plus in 40 days and nights the water would have risen at a rate of about 6 inches per minute. If that fell as rain, breathing would have been impossible because it would have been like having a fireman's hose directed into your face. There would have been no need for the flood because all breathing things would have been dead within about 15 minutes of the rain starting.

Every living substance destroyed!

And a few Bible passages need to be read and committed to memory now, because they will have a major impact on what is going to happen when the flood water runs away (where to?). They tend to be a bit repetitive, so the author obviously wanted to make sure they were clearly understood.

> And, behold, I, even I, do bring a flood of waters upon the earth, to destroy all flesh, wherein is the breath of life, from under heaven; and every thing that is in the earth shall die. (Genesis 6:17)

> For yet seven days, and I will cause it to rain upon the earth forty days and forty nights; and every living substance that I have made will I destroy from off the face of the earth. (Genesis 7:4)

> And all flesh died that moved upon the earth, both of fowl, and of cattle, and of beast, and of every creeping thing that creepeth upon the earth, and every man: All in whose nostrils was the breath of life, of all that was in the dry land, died. And every living substance was destroyed which was upon the face of the ground, both man, and cattle, and the creeping things, and the fowl of the heaven; and they were destroyed from the earth: and Noah only remained alive, and they that were with him in the ark (Genesis 7: 21-23).

Every living substance! Nothing outside the Ark remained alive.

What was floating in the water?

And so our happy band of survivors bobbed about on the sea while raging torrents of water swept unhindered around the globe, full, no doubt of the bodies of every living thing, the uprooted trees and other plants, and the suspended clays and sands that the churned up soil would have become, mingled with the decaying debris of plants and animals, while the bodies of the dead remained uncorrupted by bacteria because they would have been killed too, to settle out in a global layer of jumbled up bodies from disconnected landmasses, to fossilise in the ideal anoxic sediment for making fossils in; plants and animals all together.

Elephant from Africa; bison from North America; kangaroos from Australia with polar bears and penguins; sheep, camels, goats, pigs, and horses all jumbled up together with lions, zebras, wildebeests, aardvarks and tapirs, and fossilizing away nicely to be discovered by modern geologists.... Er... except that they're not. There is no such predictable and inevitable hydrologically sorted layer of fossil-laden silt anywhere on the surface of the entire globe.

Where is the sediment?

I expect someone must have cleaned it all up once the flood water had all run away (where to?) Well, through the Grand Canyon, down the Colorado River and into the Pacific Ocean, apparently.

What about fresh air?

So, our happy band were confined to their box, with only a single 9 square feet of window to let the carbon dioxide, methane and hydrogen sulphide out and fresh, oxygen-laden air in. Has anyone ever calculated how much oxygen a million or two animals need and how much waste gas they produce? Without a mechanical fan to circulate the air and bearing in mind that two of the three decks did not have window at all, most of the animals would have asphyxiated within a few hours if they hadn't drowned trying to breath in rain falling at 6 inched per minute.

Mind you, since the ruminants and all the other animals need billions of microbes in their gut to digest their food and maintain a healthy digestive tract, with only two (or seven) of each species of bacteria, there wouldn't have been a great deal of methane and hydrogen sulphide being produced, - and little in the way of digestion for that matter.

Who played host to the parasites and STDs?

And presumably someone or something was playing host to the intestinal parasites such as tape worms, liver flukes and round worms, along with a pair of all the other commensal, species-specific obligate parasites like lice, fleas, and the various sexually transmitted disease such as chlamydia, syphilis and gonorrhoea! Which begs the question, how did a 'righteous man' and his family acquire them in the first place?

A dove finds a living tree. Oops!

But somehow, miracle of miracles, they did, and by an even greater miracle, the dove Noah sent out to look for dry land managed to find an olive tree still alive and growing on top of a mountain! So, either God hadn't gone through with his plan to kill every living substance outside the Ark, or the story-teller either forgot that little detail or didn't realise an olive tree is a living thing – which poses a bit of a problem if you're going to try to pass of the tale as real history related by a creator god – but the Christian Church has faced bigger problems than a mere contradiction in their inerrant word of God.

What did the survivors eat?

But the survivors' troubles were now starting to get really serious, because as they struggled off the Ark, taking their first breath of fresh air in a year, they found themselves knee deep in silt, stinking of dead animals, but nothing whatever to eat!

Recall that God had thoughtfully killed every living thing!

There would have been no plants for the herbivores and once the carnivores have eaten all the herbivores they would have starved to death. The bats, shrews, hedgehogs and insectivorous birds would have quickly exterminated all the insects and other invertebrates, then starved to death, and Earth would have faced a mass extinction like no other with what few species survived facing the consequences of a narrow genetic bottleneck with populations so low that they would have been functionally extinct anyway.

And most importantly, with no green plants there would have been no oxygen for the survivors to breath!

Nothing could survive on a sterile anoxic Earth.

And that would have been that. A bare, baren, anoxic Earth with little life on it and its total human population long gone. And its ocean depths deoxygenated and lifeless from the dead and decaying sediment. It would have taken another 3 billion years for life to start up again and evolve to the level of biodiversity of about 4,000 years ago and there would have been no guarantee that anything like a sentient human being would evolve.

Just too implausible to be taken seriously.

Do you know something? This tale is beginning to sound like someone who made it up just had not thought things through and had no idea about ecology, physiology or exchange of gasses during respiration and had no idea how high the highest mountain was. They had no idea just how many species there were in the world or even that plants are living substances like breathing animals are.

Why was the tale so ludicrously implausible?

The problem with being a Canaanite pastoralist in the Bronze Age, is that your view of the world tends to be limited to how far you can walk in a day or two and what you can see from the low hills you live on. And with no science, you would have had no idea about physiology or respiration or even what 'life' is, so the tales you make up reflect that level of ignorance and scientific illiteracy.

They knew nothing about needing oxygen. But then these are the same people who saw no problem with their tale of a god creating green plants before it created the source of energy for their process of photosynthesis – the sun (Genesis 1: 12-19).

But then you could never have guessed that one day some fool was going to write your tales down and bind them up in a book and declare it to be the inerrant word of a creator god and so make that creator god look like an ignorant Canaanite pastoralist from the Bronze Age.

The sad fact is that there are people who believe those myths are real history, so fail to appreciate what they tell us about the origins of our civilisation in the fearful, superstitious, and scientifically backward infancy of our species and show us how far we have progressed since then.,

The Tall Tale of The Tower of Babel.

Soon after the Ark debacle we come across another implausible tale – the story of the Tower of Babel – presumably an attempt to explain why there are different languages.

The first problem is that the tale again got bound up in a book that it was never intended for, and the compiler was too slipshod the make sure the narrative had no wrinkles in it.

To see the blunder this caused we need to read the chapter before the tall tale of the Tower., and here we see the join between two different versions of why there are different languages:

Genesis 10 tells us that the different sons of Noah, Shem, Ham and Japheth each had children and grandchildren and great grandchildren (i.e. three generations from Noah) and the land was divided up between these great grandchildren, each speaking a different language!

Now, I have grandchildren and might in the next ten years or so, have some great grandchildren born in the UK. It would be astonishing if they spoke a different language to me, but if we are to believe the authors of this tale, it was commonplace for great grandchildren to speak a different language to the great grandparents.

Anyhow, moving on, because that is not the most important blunder. There is an even bigger one coming. But it is worth perusing that chapter again to work out how many generations there had been since Noah before the Tall Tower Tale begins. Remember, Noah, his wife and their three ancient sons and their wives were the sole survivors on Earth. No one else was there to produce children!

How many generations since Noah?

I make it three generations: four at the most. Let's be generous and assume four generations.

So, with each of four generations consisting, of about three of four sons, each speaking a different language. There would have been about $4^4 = 256$ different languages.

How many languages?

Now after reading that detailed explanation for why there were some 256 languages, turn to Chapter 11 and what is the first thing we read?

> And the whole earth was of one language, and of one speech. (Genesis 11:1)

Oops! Someone did not do his homework! Or, more likely the authors of these two tales never intended them the be consecutive chapters in the same book declared to be the inerrant word of a god, and poor editorial control did the rest!

So, let us move on because this is where it starts to get really silly. Recall that there were about four generations from Noah and let us assume their reproductive rate remained about the same, i.e. three or four sons and about the same number of daughters.

With 256 sons and their 256 wives the next generation would have been about 256 x 8 = 2048 people,

And this small handful of people were to undertake a major civil engineering project!

As with any civil engineering project the workers need a productive population to keep them supplied with the food, clothes and shelter they would otherwise be producing for themselves, so let's assume half the population could keep the other half free to build a massive tower – in other words a working man could produce enough for two people – a level of productivity unmatched today with all our technology and labour-saving devices, but let's not make things too difficult for our busy tower builders because there would only be a little over 1000 of them (512 if only the men laboured while the women stayed at home as was the norm in those days), and they have to make the tower big enough to reach above the clouds – which requires a massive base to stop it falling over.

So, problem number one is the lack of a sufficiently large population to undertake such a massive civil engineering project.

Why The Bible Cannot Be Taken Seriously

An Omniscient God Gets a Surprise

Meanwhile, what is going on up in Heaven? Because of course, in those days Heaven was just above the clouds over the Middle East, what with Earth still being flat and small with a dome over it.

Up in Heaven, so we are to believe, an omniscient god had no idea what was going on 'down there' so, as with the Adam and Eve tale, and later with the Incestuous Lot story, this omnipotent god had to 'go down' to find out what was happening, and what he found out alarmed him! An omniscient god was taken by surprise! Perhaps he hadn't become omniscient at that stage. (Genesis 11:5-7).

Problem number two: Heaven is up above what is still a small flat Earth, and an omniscient god gets taken by surprise when he 'goes down' to find out what is going on.

Then, although he is omnipotent, this god feels he needs to take immediate action to stop people cooperating, because it frightens him. And, instead of simply flicking the tower over with an almighty finger, or treading on it with a divine foot, he 'confounds their tongues' so they can no longer speak to one another. Apparently, the god who created humans didn't realise they could learn more than one language!

And it gets worse! We now have to believe the people who migrated to places like China, India, Egypt, etc, with their long cultural histories, all just happened to speak the language the people living there had spoken before the genocidal flood, or at least wrote it in the same form of writing that had been used for many centuries prior to their arrival.

And, these migrants from a culture that had witnessed the flood and had their tongues confounded, forgot all about Noah and his sons, and how their lovely big tower had been their undoing, and invented new gods and new religions to go with their old style of writing while those who remained behind remembered the whole thing, in word-perfect detail, including what God had said to himself, apparently.

Lot of Nonsense

The next Bible Tale that gets less and less plausible the more you examine it, is the story of Lot and his incest.

Again, it includes a god who hadn't yet become omniscient, apparently, and who has to 'go down' (from that Heaven that's still up from the Middle East) to find out what's going on. It is also a god who is not too fussed about the morals of someone whom he declares to be 'righteous'.

The biblical story of Lot has no obvious reason for being in the Bible, at least as far as morality tales go. In fact, it shows the god of the Hebrews in a poor light, overly obsessed with what humans do with their genitalia but having a low regard for women and no problem at all with incest. It also portrays it as far from omniscient, unsure of what justice means and easily persuaded by a mere human but then capriciously changing its mind and killing innocent people anyway simply for being there at the time.

God get a lesson in Morality.

It starts off with Abraham bargaining God down from killing everyone unless Abraham can find 50 'righteous men' in the city of Sodom, to saving them if Abraham can find just 10 such men, so giving God a lesson in justice and mercy.

But no such men are found, apparently because God then sends two angels disguised as ordinary men to the city to start the genocide.

And, as angels, obviously they need a safe house to stay in while they do the dirty deed, and God picks on Lot, 'a righteous man' in whose house they are to lodge.

But word gets out and soon a large crowd of sex-obsessed homosexuals gathers outside Lot's house demanding to have access to these pretty young men – the way homosexuals do! Soon 'all the people from every quarter' are surrounding Lot's house. Women and all? We are not told, but the storyteller might not have regarded them as people. As we shall see, he has scant respect for women.

Lot offers his two virgin daughters to a mob to be gang raped.

But Lot has a plan. He offers the crowd his two virgin daughters to be used as they wish! Yep! That is exactly what a righteous man of the times would do, apparently. Instead of defending his daughters' honour like any decent father, this righteous man offers them up to be gang-raped, to save a couple of angels who apparently lack the power to save themselves, even though they are there to kill everyone in the city (except Lot and his family who get to keep their lives as a reward for their generosity - including the daughters who have had no say in the matter, being mere women, but the story-teller needs them for something special later).

Everyone is blinded – but not everyone.

But anyway, the angels save the day by making everyone outside the house go blind! Remember that. Everyone, from every quarter (except Lot and his daughters) are now blind.

Next, the angels tell Lot to go and gather up his family – his married daughters and their husbands. Apparently, they aren't part of the 'every person from every quarter' who are now blind. But they think he is joking. All around went blind last night but they think their father-in-law has gone do-lally-tap when he tells them something even nastier is about to happen.

So, Lot sets out with his wife and two daughters while God rains down fire and brimstone on the city, destroying everyone in it in another one of his favourite war crimes – genocide.

Lot's wife achieves fame in her own right.

But, alas for Lot's wife, who has gone down in history as merely the appendage to her famous husband, and has no name if her own, she turns to look back at the death and destruction going on behind her and this make God so mad he turns her into a pillar of salt! And newly widowed Lot shrugs, and he his two remaining daughters, the virgins whom he offered up to be gang raped, trudge on and eventually find a cave to live it.

Lot gets drunk – but not THAT drunk.

Now, the daughters are getting a bit fed up with being virgins, so, they get some wine from somewhere (presumably from down the local cash and carry) and give so much to their father that he becomes legless and insensible, but still able to rise to the occasion, when his daughters take it in turn to date-rape him.

And so, they both end up pregnant as the result of non-consensual sex on the part of the father (theirs and their babies'). One of their babies is called Moab who went on to be the progenitor of the Moabite people. The other is called Benammi who was to be the progenitor of the Ammon people.

What it was really all about.

And that is the whole point of this unlikely tale! To produce two local related tribes of people, the Moabites and the Ammonites, the product of incest!

At least that's the conclusion of two Israeli archaeologists from Tel Aviv University, Israel Finkelstein and Neil Asher who, in their book *The Bible Unearthed: Archaeology's New Vision of Ancient Israel and the Origin of Sacred Texts*, (102) explain how this only makes any sense if this tale was made up in the 7th-century when it would have had any relevance given the complex political relationships at that time.

> The relationships of Israel and Judah with their eastern neighbors are also clearly reflected in the patriarchal

narratives. Through the eighth and seventh centuries BCE their contacts with the kingdoms of Ammon and Moab had often been hostile; Israel, in fact, dominated Moab in the early ninth century BCE. It is therefore highly significant — and amusing—how the neighbors to the east are disparaged in the patriarchal genealogies. Genesis 19:30–38 (significantly, a J text [1]) informs us that those nations were born from an incestuous union. After God overthrew the cities of Sodom and Gomorrah, Lot and his two daughters sought shelter in a cave in the hills. The daughters, unable to find proper husbands in their isolated situation— and desperate to have children— served wine to their father until he became drunk. They then lay with him and eventually gave birth to two sons: Moab and Ammon. No seventh century Judahite looking across the Dead Sea toward the rival kingdoms would have been able to suppress a smile of contempt at a story of such a disreputable ancestry.

The Fishy Tale of Jonah

Having dealt with some of the more implausible tales in Genesis that unfortunate creationists are obliged to try to defend as literal truth and real history, we can move on to another story that doesn't seem to serve any purpose as a morality tale or an allegory for anything, other than as a reminder of God's genocidal tendencies. This one gets a book all to itself and appears to have been added for mere padding.

---

[1] J texts are Bible stories which mostly refer to God as Yahweh (YHWH) and attributed to the Yahwists (Jahwist in German) as opposed to E text stories written by the Elohists which use the name Elohim or El. The J texts are mostly associated with the southern kingdom of Judah while the E texts tend to be associated with the northern kingdom of Israel. The two strands of stories and 'histories' have later been woven together into a single narrative, often not very successfully.

It is the tale of Jonah and how he spent three days living inside the stomach of a great fish (or was it a whale? 'Inerrant' versions differ on that point.)

Jonah is given a job to do.

It starts off with God telling Jonah he has a little job for him. He is to go to the city of Ninevah and tell the ruler and the townsfolk that God's about to destroy the lot of them for being 'wicked' (no further details on that point).

Jonah tried to flee to Tarshish.

Now Jonah, understandably perhaps, doesn't want to take that risk because you know what tyrannical rulers do to messengers who bring bad news, so he reckons he can run away and hide from the omniscient god who has the powers to destroy a city at the flick of divine digit, so he goes the Joppa and gets on a boat to Tarshish.

Sailors believe Yahweh is out to get Jonah.

Then weird things happen. The boat is hit by a storm, and the sailors who all believe in different gods, conclude that it must be Yahweh because Jonah is on the boat, so they throw him overboard. Apparently, God can cause a wind to threaten a boat but can't arrange for it to turn a sailing ship round and head back to Joppa.

A Home from home in the Stomach of a fish.

And just as Jonah must have been thinking things could have get any worse as he begins to go down for the third time, things get worse; he become the dinner of a 'great fish', where he lives for three days, unaffected by the fishes' digestive enzymes and managing without food, water or oxygen. A normal person would die for want of water in three days and would have died of asphyxia in a few minutes whereupon the digestive enzymes would be dissolving the remaining bones within three days, but not Jonah.

The transgender fish gets sick.

Two miracles, apart from his remarkable survival are about to happen. Firstly the 'great fish' which in the Hebrew version of the story was a *dag gado* (great male fish) magically turns into a female and vomits Jonah up unscathed and none the worse for his NDE. In the Hebrew version the fish has changed sex and become a *daga* (female fish) when it deposited Jonah onto the beach.

God tries again.

"Okay!" says God, "You've seen what I can do if you don't do what I want! Nowe how about that little job I gave you?

And Jonah accepted the offer he could not refuse and set off for Ninevah.

As an aside from this tale: It's been something of an embarrassment for Christians since the days of Augustine of Hippo who complains of being laughed at by 'pagans' when he tried to tell them the tale is true.

> "The last question proposed is concerning Jonah, and it is put as if it were not from Porphyry, but as being a standing subject of ridicule among the Pagans; for his words are: "In the next place, what are we to believe concerning Jonah, who is said to have been three days in a whale's belly? The thing is utterly improbable and incredible, that a man swallowed with his clothes on should have existed in the inside of a fish. If, however, the story is figurative, be pleased to explain it. Again, what is meant by the story that a gourd sprang up above the head of Jonah after he was vomited by the fish? What was the cause of this gourd's growth?" Questions

such as these I have seen discussed by Pagans amidst loud laughter, and with great scorn. (103)

Jonah warns the people of Ninevah

Now picture the scene: A dishevelled character, smelling of fish, turns up in the town square and starts ranting about sinners and how a voice had told him God will destroy them and their city in 40 days unless they repent and change their wicked ways. He tells them he has just spent three days in the belly of a fish, so he knows it is true.

And they believed him!

Has the king taken leave of his senses?

So did the king of Nineveh, apparently! He promptly took his clothes off, dressed in sackcloth, sat in ashes, and ordered that none of the townsfolk or their livestock were to eat or drink anything, but they were all to wear sackcloth.

Anyway, it worked. Not only did the people not rise up and overthrow their obviously derange king, but God did not destroy the city.

That reminds me of a story I once heard of a man on a train in England who every few minutes stood up and threw a paper tissue out of the window. When asked why, he explained that it stopped the rhinoceroses from attacking the train. "But there are no rhinoceroses in England!" said a fellow passenger.

"Well, that proves it works!" said the man triumphantly.

To recap then, this dishevelled character has told the people and their king to wear sackcloth and ashes to stop their city from being destroyed. The city is not destroyed. That proves it worked, apparently.

Jonah throws a tantrum because Nineveh is not destroyed!

Now we are to believe that having gone to all that trouble to warn the people of Nineveh to mend their ways or be destroyed, and having persuaded them to mend their ways, Jonah goes into a big sulk because the city wasn't destroyed!

He complains to the voice in his head that life is not worth living anymore!

The Magic gourd.

So, he sits outside the city and sulks, but God takes pity on him and magically makes a gourd (a sort of pumpkin) grow up and make a shelter for him to sit under, where he could sit and wait to see the city destroyed after all.

Then, for reasons known only to God, he created a 'worm' that destroyed the gourd. and left poor Jonah, still sulking and depressed under the blazing sun with no shelter, still waiting to see the city destroyed but now feeling sorry for the gourd that had done nothing to deserve its fate.

So, says God, you feel sorry for the gourd and wish I had spared it, yet you think I should destroy a city with all its people! Don't you think you need to reassess your priorities?

"Fair enough!" says Jonah, "Point taken." And we hear no more of him.

And we are also expected to believe that either Jonah wrote this tale himself, or he told it to some other person who thought it was true and wrote it down, whereupon it got bound into a book along with other implausible tales and proclaimed to be the inerrant word of an omnipotent creator god.

And it you believe that you will probably believe anything and will give money to whomsoever to you.

Would You Adam and Eve It?

The final piece of Bible absurdity is one that deserves a chapter all to itself. It is not only fundamental to creationism but to every other Christian sect, to Judaism and to Islam, although the more moderate sects tend to treat it as an allegory or metaphor without explaining for what.

It is also interesting for what it tells us of the primitive theologies that coalesced to form Christianity, because it makes more sense as a Gnostic allegory than a Judaeo-Christian text.

It is, of course the tale of Adam and Eve and of how their 'original sin' was deemed by God to be inherited by each and every one of us, so creating a cradle to grave sin for which we need to atone and seek redemption. This is not the act of an all-loving, merciful god but of an insecure control freak who lacks confidence in his ability to earn love and respect so needs to command it – much like a Bronze Age tribal despot who is always at risk from being overthrown in an extremely messy way.

The idea of original sin, and an unspeakably horrible punishment awaiting us if we do not fully atone for it, was a masterstroke for the priesthood, because it gives them control over us that no religion had ever managed to achieve before. Like A person with Munchausen by Proxy Syndrome, they make us mortally sick and take credit for providing the cure, and we are obliged to admire their wisdom and expertise and give them power over all aspects of our lives.

So, what is this tale and why is it a Gnostic allegory and what does it tell us of the god as it was perceived by the authors in the fearful infancy of our species?

The magical creation of Adam

It starts off with God magically creating a man, Adam, out of dust by breathing life into him. In those days, obviously, beathing and living were synonymous as we saw in the Noah's Ark tale where every living thing being killed probably meant every breathing thing so didn't include the plants, which weren't regarded as living things.

We then go through the second version of creation, the first one being where animals were created before humans, to the second version where the animals were created for the benefit of Adam, but none of them were suitable. Apparently, this omniscient god didn't know what animals would be suitable and ended up creating none that were, so all the animals are there for our use, and are the mere detritus of a failure of design.

The cloning of Eve.

So, he decided to create Eve, who, although she was biologically a clone of |Adam, had a female set of sex chromosomes – not a problem if you're a Bronze Age storyteller with no knowledge of genetics making up stories an equally ignorant audience.

And now this is where the Gnostic stuff starts to creep in, so let's consider the Gnostics.

The Gnostics

They were an early religious sect that believed the whole of the physical realm was the creation of an evil god who created it as a distraction to create a wedge between humans and the spiritual realm where God lives. Gnosticism played a major role in the development of the Cathar religion that became predominant in Southern France in the early Middle Ages.

The Cathars claimed to be Christian but refused to acknowledge the supremacy of the Pope and, even more seriously, refused to pay tithes, so, in a demonstration of Christian love and tolerance for their fellow man, the Cathars were brutally suppressed and massacred by Catholic Christians in the so-called Albigensian Crusade (the Bishop of Albi in Southern France had converted to Catharism, taking his entire flock with him). Massacres were perpetrated with the enthusiastic encouragement of the inappropriately named Pope Innocent, most notably at .Bezier and Carcassonne as I relate in my blog posts on the subject (104) (105).

There are unwitting echoes of Gnosticism in creationism where the physical record is normally regarded as a lie created by Satan and the scientists who discover it are agents of Satan trying to trick us into doubting God's inerrant word in the Bible.

So, back to our tale of Adam and his new playmate, Eve.

Forbidden fruits and God's lie to Adam

The story goes that God created a garden for them to live in full of fruit and other comestible but there was one tree they were forbidden to eat from because, as God warned Adam, if they did, they would 'die that day'. The tree is the tree of knowledge and to eat from the fruits of it is to gain the knowledge of good and evil (something that was the preserve of gods in those days, apparently). Why God put it there is never explained.

Some interpret 'eating of the tree of knowledge' as having sex, which God had forbidden even though he had created them with fully functional genitalia. But to a creationist, there was a literal tree with literal poison fruit because that's what it says in the Bible. Apparently, Jesus was the only character in the Bible who used metaphors and parables; everyone else spoke the literal, inerrant truth.

The serpent tells Eve the truth.

Then along comes a serpent (a sexual metaphor?) and tells Eve that's a lot of nonsense and it is safe to eat the fruit of the sacred tree. Eve tells Adam and Adam believes her, and they share an apple (sexual metaphor again!)

And they don't die that day!

Turns out, God had lied Adam! In his first words to Adam god lied, and the serpent told Eve the truth. They didn't 'die that day'; instead, they gained 'the knowledge' or 'gnosis.

In the Gnostic tradition, of course, the creator of the physical world is the evil one, and here we have Adam's creator lying to him to prevent humans gaining 'the knowledge', or Gnosis, which is the knowledge of how to get to the spiritual realm where the real God lives.

God learns the truth.

Then God, who in his pre-omniscient stage lived in the Heaven that was 'up' above the Middle East, (something that only makes any sense on a small, flat Earth) had to 'go down' to find out what was going on and discovered that the couple are now hiding (from an omnipotent god!) and when he finds them he discovered they have covered their genitalia with fig leaves! There were only two of them.so they had nothing to hide that they hadn't seen before! But from that God concludes that they must have eaten the forbidden fruit and now know right from wrong.

The divine set-up.

Of course, before they gained that knowledge, they wouldn't have known that disobedience was wrong, but God deems them to have sinned anyway. He had fixed it, so they didn't know right from wrong and had tempted them with some tasty-looking fruit before they knew it was wrong to disobey!

It was clearly a set-up, but he decides to impose harsh penalties on them for the 'Sin' that their children and their children's children and so *ad infinitum* would also inherit. They are to carry the burden of sin for all their lives, and God was so incensed by their disobedience, he would never be able to forgive them for it.

Men and women are allotted their roles in life. That'll teach 'em!

Now comes the bits that look exactly like the later requirements of a ruling class – men are to toil for a living and women are to be the possession of men, a slave to the whims of their menfolk, and are to suffer in childbirth as a constant reminder of what Eve did. How convenient!

Have you guessed who wrote it yet?

How nicely that all turned out for the ruling male tribal leadership, eh? An excuse for coercive control and all to be blamed on a god because of something some remote ancestor did.

And there we have the beginnings of religion providing excuses for people who need excuses, and why the ruling classes have been hand in glove with the priesthood ever since.

Conclusion

That creationism is a political movement is beyond doubt since the Discovery Institute's Wedge Strategy has subverting the US Constitution and overturning the Establishment Clause as a major strategic goal. The ultimate aim is to dismantle Thomas Jefferson's 'Wall of Separation' between Church and State, abolish the secular constitution and replace it with a fundamentalist Christian theocracy.

In pursuit of that objective, creationists exploit the ignorance and religious fundamentalism of their target marks with disinformation, with no regard for intellectual honesty and integrity and a clear intention to mislead rather than inform.

One major strategic aim is to link the political objective with religious fundamentalism by trying to persuade Bible-believing Christians that the opening chapter of the Bible, Genesis, is literal science and history with solid evidential support. The objective is to harness the religious culture in American society and persuade Christians that the political objectives of the Discovery Institute is a religious crusade to overthrow the 'evils' of science and scientists.

Another important strategic aim is to forge links with conservative Christian groups, the success of which we can see in the creationist infiltration of the Republican Party which has become the political wing of Evangelical Christianity in alliance with creationist groups.

That their real objective is political rather than religious can be seen in their readiness to ignore the fundamental principles of the religions they are using as an excuse, especially the prohibition in Christianity on bearing false witness.

I have shown how lies and misrepresentations of science and the accusations of falsification and deliberate misinterpretation of data by scientists is the standard fare of the creationist disinformation strategy with the stated objective of destroying public trust in science and scientists.

What this tactic shows us is that they don't believe science is wrong; they know science is right, but they need people to believe otherwise because they know that science undermines and refutes their religious agenda.

In this book I have shown that the traditional claims of the science-denying creationists are wrong and demonstrably so. There are scientific explanations for events and processes such as the Big Bang, abiogenesis, new genetic information and, 'macro-evolution', with abundant fossils showing transition over time from earlier forms to more recent forms.

Multiple strands of evidence from the fossil record, through genetics and comparative anatomy and physiology converge on a pattern of nested hierarchies and a branching evolutionary tree showing the relationships between extant and extinct species, leaving no room for doubt that the present biodiversity is the result of evolutionary processes over time.

I have also shown that creationists have a fundamentally false idea of what evolutionary scientists define as evolution and especially 'macro-evolution', and how these false ideas are deliberately promulgated by creationist cult leaders as straw men to attack.

I have shown examples of deliberate deception using quote mines and misrepresentation to create the idea that there is scientific support for creationism where there is none.

I have also shown why the scientific evidence shows that Earth is billions of years old in an even older Universe and that the Bible description of *ex nihilo* creation just a few thousand years ago is wrong, and can not be explained as a metaphor or an

Conclusion

allegory, but is fundamentally wrong and based on the stories of superstitious Bronze Age pastoralists with a limited and parochial view of the world and no science on which to base their tales.

They lived, in Christopher Hitchin's words, in the "fearful infancy of our species" in which the world ran on magic, and spirits and demons were responsible for natural phenomena from rain, thunder and lightning to diseases, plagues and earthquakes. A society ruled by tribal despots in which women and slaves had the same status as cattle.

There is also evidence that their tales were never intended to be collected up and bound together in a book declared to be the inerrant word of an omniscient creator god.

I have also shown why the Bible stories themselves refute the idea that they are real history related by an omniscient creator god. Quite simply, the Bible could not have been written of inspired by the god described in it.

References

1. *Creationism and conspiracism share a common teleological bias.* **Wagner-Egger, Pascal, et al.** 16, s.l. : Elsivier, 2018, Current Biology, Vol. 28, pp. R867-R868. 0960-9822.

2. *Intuitive biological thought: Developmental changes and effects of biology education in late adolescence.* **Coley, John D., et al.** 1, s.l. : Elsevier, 2017, Cognitive Psychology, Vol. 92. 0010-0285.

3. *Professional physical scientists display tenacious teleological tendencies: Purpose-based reasoning as a cognitive default.* **Kelemen, D., Rottman, J., & Seston, R.** 4, s.l. : American Psychological Association, 2012, Journal of Experimental Psychology: General, Vol. 142, pp. 1074–1083.

4. **Menton, Dr. David.** 10 Best Evidences From Science That Confirm a Young Earth #3 Soft Tissue in Fossils. *Answers in Genesis.* [Online] 1 October 2012. [Cited: 19 October 2024.] https://answersingenesis.org/fossils/3-soft-tissue-in-fossils/.

5. **Catchpoole, David & Sarfati, Jonathan.** 'Schweitzer's dangerous discovery'. *Creation.com.* [Online] [Cited: 19 October 2024.] https://creation.com/schweitzers-dangerous-discovery.

6. **Bucannan, Scott.** Letters to Creationists. *Dinosaur Soft Tissue.* [Online] [Cited: 19 October 2024.] https://letterstocreationists.wordpress.com/dinosaur-soft-tissue/.

7. *Soft Tissue Preservation in Terrestrial Mesozoic Vertebrates.* **Schweitzer, Mary Higby.** s.l. : Annual Reviews, 2011, Annual Review of Earth and Planetary Sciences, Vol. 39, pp. 187-216.

8. **Idle, Eric and Du Prez, John.** Galaxy Song. *LyricFind.com.* [Online] 1989. [Cited: 19 October 2024.] https://lyrics.lyricfind.com/lyrics/monty-python-galaxy-song.

9. **National Center for Science Education.** The Wedge Document. *NCSE.ngo*. [Online] National Center for Science Education, 14 October 2008. [Cited: 5 November 2024.] https://ncse.ngo/wedge-document.

10. **Discovery Institute.** The "Wedge Document" so what? *Discovery.org*. [Online] April 2019. [Cited: 5 November 2024.]

11. **Behe, Michael J.** *Darwin's Black Box: The Biochemical Challenge to Evolution.* 2nd. s.l. : Free Press, 2006.

12. **Brenan, Megan.** Majority Still Credits God for Humankind, but Not Creationism. *Gallup.* [Online] Gallup Inc., 2 July 2024. [Cited: 2 November 2024.] https://news.gallup.com/poll/647594/majority-credits-god-humankind-not-creationism.aspx.

13. **Rubisondior, Rosa.** *The Unintelligent Designer: Refuting the Intelligent Design Hoax.* s.l. : CreateSpace Independent Publishing Platform, 2018. ISBN-13 : 978-1723144219.

14. **Rubicondior, Rosa.** *The Malevolent Designer: Why Nature's God is Not Good.* s.l. : Independently published (Amazon), 2020. ISBN-13 : 979-8670361729.

15. **Behe, Michael J.** *The Edge of Evolution: The Search for the Limits of Darwinism.* s.l. : Free Press, 2007. ISBN-13 : 978-0743296205.

16. *Diverse mutational pathways converge on saturable chloroquine transport via the malaria parasite's chloroquine resistance transporter.* **Summers, Robert L., Dave, Anurag and Tegan J. Dolstra, et al.** 17, s.l. : National Academy of Sciences, 11 April 2014, Proceedings of the National Academy of Sciences (PNAS), Vol. 111.

17. **Behe, Michael J.** *Darwin Devolves: The New Science About DNA That Challenges Evolution.* s.l. : HarperOne, 2020. ISBN-13 : 978-0062842664.

References

18. **Rubicondior, Rosa.** Creationism In Crisis - How We Know Earth Is 4.5 Billion Years Old. *Rosa Rubicondior.* [Online] 24 June 2024. [Cited: 19 October 2024.] https://rosarubicondior.blogspot.com/2024/06/creationism-in-crisis-how-we-know-earth.html#zircon.

19. *Can We Get a Good Radiocarbon Age from "Bad Bone"? Determining the Reliability of Radiocarbon Age from Bioapatite.* **Cherkinsky, Alexander.** 2, s.l. : ResearchGate, 2009, Radiocarbon, Vol. 51, pp. 647–655 .

20. **Darwin, Charles and Russell, Alfred.** *The Original Papers by Charles Darwin and Alfred Russell Read To The Linnean Society, London.* s.l. : Univrsity of British Columbia, 1858. PDF.

21. **Darwin, Charles.** *On the Origin of Species By Means Of Natural Selection, Or The Preservation of Favoured Races in the Struggle For Life.* 1. London : The Natural History Museum, 1859. 978-0565095024.

22. *Local cryptic diversity in salinity adaptation mechanisms in the wild outcrossing Brassica fruticulosa.* **Busoms, Silvia, da Silva, Ana C. and Escolà, Glòria & Yant, Levi .** [ed.] Jonathan Wendel. 40, s.l. : National Academy of Sciences, 24 September 2024, Proceedings of the National Academy of Science (PNAS), Vol. 121.

23. **Public Broadcasting Service.** Genetic Drift and the Founder Effect. *PBS.org Evolution.* [Online] Public Broadcasting Service. [Cited: 20 October 2024.] https://www.pbs.org/wgbh/evolution/library/06/3/l_063_03.html .

24. **Anon.** Cheetahs: On the Brink of Extinction, Again. *National Geographid Education.* [Online] National Geographic. [Cited: 20 October 2024.] https://education.nationalgeographic.org/resource/cheetahs-brink-extinction-again/5th-grade/.

25. *Effective population size and evolutionary dynamics in outbred laboratory populations of Drosophila.* **Mueller, Laurence D, et al.** s.l. : Springer Nature Ltd., 2013, Journal of Genetics, Vol. 92, pp. 349–361.

26. *A Draft Sequence of the Neandertal Genome.* **Green, Richard E., et al.** 5979, s.l. : American Association for the Advancement of Science., 7 May 2010, Science, Vol. 328, pp. 710-722.

27. *Genetic evidence for complex speciation of humans and chimpanzees.* **Patterson, Nick, et al.** s.l. : Springer Nature Ltd., 17 May 2006, Nature, Vol. 441, pp. 1103–1108.

28. *Clonal genome evolution and rapid invasive spread of the marbled crayfish.* **Gutekunst, Julia, et al.** s.l. : Springer Nature Ltd, 5 February 2018, Nature Ecology & Evolution, Vol. 2, pp. pages567–573.

29. *Phylogenetic Relationships of Whiptail Lizards of the Genus Cnemidophorus (Squamata: Teiidae): A Test of Monophyly, Reevaluation of Karyotypic Evolution, and Review of Hybrid Origins.* **Reeder, Tod W., Cole, Charles J and Dessauer, Herbert C.** 3365, s.l. : BioOne, 17 May 2002, American Museum Novitates, Vol. 2002, pp. 1-61.

30. *Accumulation of endosymbiont genomes in an insect autosome followed by endosymbiont replacement.* **Tvedte, Eric S., et al.** 12, s.l. : Cell Press, 20 June 2022, Current Biology, Vol. 32, pp. 2786-2795.e5.

31. *Horizontal gene transfer of the algal nuclear gene psbO to the photosynthetic sea slug Elysia chlorotica.* **Rumpho, Mary E., et al.** 46, s.l. : PNAS, 18 November 2008, Proceedings of the National Academy of Science (PNAS), Vol. 105, pp. 17867-17871.

References

32. *Convergent horizontal gene transfer and cross-talk of mobile nucleic acids in parasitic plants.* **Yang, Zhenzhen, et al.** s.l. : Springer Nature Ltd, 22 July 2019, Vol. 5, pp. 991–1001.

33. **Wildlife Insights.** Introduction to British Vanessid Butterflies. *Wildlife Insight.* [Online] [Cited: 21 October 2024.] https://www.wildlifeinsight.com/vanessid-butterflies-nymphalinae/.

34. *Recent speciation of Capsella rubella from Capsella grandiflora, associated with loss of self-incompatibility and an extreme bottleneck.* **Guo , Ya-Long , et al.** 13, s.l. : PNAS, 31 March 2009, Proceedings of the National Academy of Science (PNAS), Vol. 106.

35. *Early cave art and ancient DNA record the origin of European bison.* **Soubrier, Julien , et al.** s.l. : Springer Nature Ltd, 18 October 2016, Nature Communications , Vol. 7.

36. **Cooper, Alan and Subrier, Julien.** How we discovered the 'Higgs bison', hiding in plain sight in ancient cave art. *The Conversation.* [Online] The Conversation, 18 October 2016. [Cited: 30 October 2024.] https://theconversation.com/how-we-discovered-the-higgs-bison-hiding-in-plain-sight-in-ancient-cave-art-67231.

37. **Bohlender R, et al.** Ancient Human History More Complex than Previously Thought, Researchers Say. *American Socity of Human Genetice (ASHG).* [Online] 20 October 2016. [Cited: 22 October 2024.] https://www.ashg.org/publications-news/press-releases/201610-admixture/.

38. *Genome of a middle Holocene hunter-gatherer from Wallacea.* **Carlhoff, Selina, et al.** s.l. : Springer Nature Ltd, 25 August 2021, Nature, Vol. 596, pp. 543–547.

39. *Developmental Basis of Phallus Reduction during Bird Evolution.* **Herrera, Ana M. , et al.** 12, s.l. : Elsevier Ltd., 17 June 2013, Current Biology, Vol. 23, pp. 1065-1074.

40. *Rapid large-scale evolutionary divergence in morphology and performance associated with exploitation of a different dietary resource.* **Herrel, Anthony, et al.** 12, s.l. : PNAS, 25 March 2008, Proceedings of the National Academy of Sciences (PNAS), Vol. 105, pp. 4792-4795.

41. *The red queen in the corn: agricultural weeds as models of rapid adaptive evolution.* **Vigueira, C. C., Olsen, K. M and Caicedo, A. L.** s.l. : Springer Nature Ltd, 28 November 2012, Heredity, Vol. 110, pp. 303–311.

42. **Rubicondior, Rosa.** A Golden Case Of Rapid Evolution. *Rosa Rubicondior.* [Online] 28 January 2013. [Cited: 23 October 2024.] https://rosarubicondior.blogspot.com/2013/01/a-golden-case-of-recent-rapid-evolution.html.

43. —. Something Fishy About Creationism. *Rosa Rubicondior.* [Online] 13 January 2012. [Cited: 23 October 2024.] https://rosarubicondior.blogspot.com/2012/01/something-fishy-about-creationism.html.

44. **Peterson, Greg.** Debating the fastest evolution on record. *GeoTimes.* [Online] AgiWeb.org, 23 April 2013. [Cited: 23 October 24.] https://www.agiweb.org/geotimes/apr03/WebExtra042503.html.

45. *Culex pipiens in London Underground tunnels: differentiation between surface and subterranean populations.* **Byrne, Katherine & Nicholls, Richard A.** s.l. : Springer Nature Ltd, 1 January 1999, Heredity, Vol. 82, pp. 7-15.

46. *Natural hybridization in heliconiine butterflies: the species boundary as a continuum.* **Mallet, James, et al.** 28, s.l. : Springer Nature Ltd, 23 February 2007, BMC Ecology and Evolution, Vol. 7.

47. *Natural selection and sympatric divergence in the apple maggot Rhagoletis pomonella.* **Filchak, K., Roethele, J. and &**

References

Feder, J. s.l. : Springer Nature, 12 October 2000, Vol. 407, pp. 739–742.

48. *Hand and foot morphology maps invasion of terrestrial environments by pterosaurs in the mid-Mesozoic.* **Smyth, Robert S.H., et al.** s.l. : Elsevier Inc., 4 October 2024.

49. *Organ systems of a Cambrian euarthropod larva.* **Smith, Martin R., et al.** s.l. : Springer Nature Ltd, 31 July 2024, Nature, Vol. 633, pp. 120–126.

50. **Hecht, Jeff.** Evolution's detectives: Closing in on missing links. *New Scientist.* [Online] New Scientist Ltd, 13 February 2013. [Cited: 24 October 2024.] https://www.newscientist.com/article/mg21729041-900-evolutions-detectives-closing-in-on-missing-links/.

51. *The feeding system of Tiktaalik roseae: an intermediate between suction feeding and biting.* **Lemberg, Justin B., Daeschler, Edward B. and Shubin, Neil H.** 7, s.l. : PNAS, 1 February 2021, Proceedings of the National Academy of Science, Vol. 118.

52. *Feeding kinematics and morphology of the alligator gar (Atractosteus spatula, Lacépède, 1803).* **Lemberg, Justin B., Shubin, Neil H and Westneat, Mark W.** s.l. : John Wiley & Sons, 6 August 2019, Journal of Morphology, Vol. 280, pp. 1548–1570.

53. *The Liexi fauna: a new Lagerstätte from the Lower Ordovician of South China.* **Fang, Xiang , et al.** s.l. : The Royal Society, 13 July 2022, Proceedings of the Royal Society B, Vol. 289.

54. *Crab in amber reveals an early colonization of nonmarine environments during the Cretaceous.* **Luque, Javier, et al.** 43, s.l. : American Association for the Advancement of Science., 20 October 2021, Science Advances, Vol. 7.

55. *Earliest giant panda false thumb suggests conflicting demands for locomotion and feeding.* **Wang, Xiaoming, et al.** s.l. : Springer Nature Ltd., 30 June 2022, Vol. 12.

56. *Cretaceous arachnid Chimerarachne yingi gen. et sp. nov. illuminates spider origins.* **Wang, B., Dunlop, J.A. and Selden, P.A., et al.** s.l. : Springer Nature Ltd, 5 February 2018, Nature Ecology & Evolution, Vol. 2, pp. 614–622.

57. *Origin of spiders and their spinning organs illuminated by mid-Cretaceous amber fossils.* **Huang, D., et al.** s.l. : Springer Nature Ltd., 5 February 2018, Nature Ecology & Evolution, Vol. 2, pp. 623–627.

58. *Australopithecus sediba: A New Species of Homo-Like Australopith from South Africa.* **Berger, Lee R., et al.** 5979, s.l. : American Ossociation for the Advancement of Science, 9 April 2010, Science, Vol. 329, pp. 195-204.

59. *Australopithecus sediba Hand Demonstrates Mosaic Evolution of Locomotor and Manipulative Abilities.* **Kivell, Tracy L., et al.** 6048, s.l. : American association for the Advancement of Science, 9 September 2011, Science, Vol. 333, pp. 1411-1417.

60. *Perimortem fractures in Lucy suggest mortality from fall out of tall tree.* **Kappelman, John , et al.** 29 August 2016, Nature, Vol. 537, pp. 503–507.

61. **Smithsonian Museum.** Laetoli Footprint Trails. *humanorigins.si.edu.* [Online] Sithsonian Museum, 8 July 2024. [Cited: 26 October 2024.] https://humanorigins.si.edu/evidence/behavior/footprints/laetoli-footprint-trails.

62. **Various.** The Casimir Effect. *Wikipedia.* [Online] Wikipedia. [Cited: 30 October 2024.] https://en.wikipedia.org/wiki/Casimir_effect.

References

63. **Rubicondior, Rosa.** *What Makes You So Special: From the Big Bang To You.* s.l. : CreateSpace Independent Publishing Platform (Amazon), 2017. ISBN-13 : 978-1546788294.

64. **LePage, Michael and Lane, Nick.** How life evolved: 10 steps to the first cells. *New Scientist.* 14 October 2009.

65. **Ackerman, Paul D.** *It's a Young World After All.* Grand Rapids : Baker Books, 1986. ISBN 0-8010-0204-4.

66. **Answers in Genesis (Ken Ham, et al).** Answers in Genesis. *Statement of \Faith.* [Online] AiG, 5 March 2021. [Cited: 26 October 2024.] https://answersingenesis.org/about/faith/.

67. **Wakefield, J. Richard.** The Geology of Gentry's "Tiny Mystery". *University of Northridge - MYRMEKITE AND METASOMATIC GRANITE, ISSN 1526-5757.* [Online] University of Northridge, 26 March 1996. [Cited: 26 October 2024.] http://www.csun.edu/~vcgeo005/gentry/tiny.htm.

68. **Rubicondior, Rosa.** It's An Old World After All. *Rosa Rubicondior.* [Online] [Cited: 26 October 2024.] https://rosarubicondior.blogspot.com/p/its-old-world-after-all.html.

69. **Creationism.org.** It's a Young World After All. *creationism.org.* [Online] [Cited: 26 October 2024.] https://creationism.org/books/ackerman/index.htm.

70. **Catchpole, David.** Angkor saw a stegosaur? *Creationministries.com.* [Online] Creation Ministries International, September 2007. [Cited: 27 Octoeber 2024.] https://creation.com/angkor-saw-a-stegosaur.

71. **O'Brien, Johnathan.** A Mountain in a Year. *creationministries.com.* [Online] Creation Ministries International, January 2012. [Cited: 27 October 2024.] https://creation.com/mount-paricutin.

72. **O'Brien, Jonathan.** Radiometric backflip. *Creationminestries.com.* [Online] Creation Ministries International, 26-28 January 2018. [Cited: 27 October 2024.] https://creation.com/radiometric-backflip.

73. *Retraction Note: Bird-like fossil footprints from the Late Triassic.* **Melchor, R., de Valais, S. and & Genise, J.** s.l. : Springer Nature Lts, 7 August 2013, Nature, p. 262.

74. *Bird-like fossil footprints from the Late Triassic.* **Melchor, Ricardo N., de Valais, Silvina and Genise , Jorge F.** s.l. : Springer Nature Ltd, 27 June 2002, Nature, Vol. 417, pp. 936–938.

75. *Geological setting and paleomagnetism of the Eocene red beds of Laguna Brava Formation (Quebrada Santo Domingo, northwestern Argentina).* **H. Vizán, S. Geuna, R. Melchor, E.S. Bellosi, S.L. Lagorio, C. Vásquez, M.S. Japas, G. Ré, M. Do Campo.** s.l. : Elsevier B.V., 1 November 2013, Tectonophysics , Vol. 583, pp. 105–123.

76. **Science Direct.** Australopithecus Afarensis. *Science Direct.* [Online] Elsevier. [Cited: 27 October 2024.] https://www.sciencedirect.com/topics/psychology/australopithecus-afarensis.

77. *Australopithecus sediba and the emergence of Homo: Questionable evidence from the cranium of the juvenile holotype MH 1.* **Kimbel, William H. and Rak, Yoel.** s.l. : Elsevier, June 2017, Vol. 107, pp. 94-106.

78. **Rubicondior, Rosa.** Nebraska Man - A Creationist Hoax. *Rosa Rubicondior.* [Online] 5 February 2012. [Cited: 28 October 2024.] https://rosarubicondior.blogspot.com/2012/02/nebraska-man-creationist-hoax.html.

79. —. Piltdown Man - A Triumph For Science. *Rosa Rubicondior.* [Online] 5 February 2012. [Cited: 28 October

References

2024.] https://rosarubicondior.blogspot.com/2012/02/piltdown-triumph-for-science.html.

80. **Frank, Spencer.** *The Piltdown Papers, 1908-55: The Correspondence and Other Documents Relating to the Piltdown Forgery.* s.l. : Oxford University Press, 1990. ISBN-13 : 978-0198585237.

81. *I.—Note on the Piltdown Man (Eoanthropus Dawsoni).* **Woodward, A. Smith.** 10, October 1913, Geological Magazine, Vol. 10, pp. 433 - 434.

82. **Webb, Johnathan.** *Piltdown review points decisive finger at forger Dawson.* Science & Environment, BBC News. s.l. : BBC, 2016. New report.

83. **Time Magazine.** *Science : End As a Man.* s.l. : Time Magazine (WaybackMachine Archive), 1953. News report.

84. **The Talk Origins Archive.** *The Quote Mine Project. The Talk Origins Archive.* [Online] Talk Origins, 2003. [Cited: 28 October 2024.] https://www.talkorigins.org/faqs/quotes/mine/part1-1.html.

85. **Ayala, F. J. and Valentine, J. W.** *Evolving: The Theory and Process of Organic Evolution.* 1978.

86. **Futuyama, Douglas.** *Science on Trial: The Case for Evolution.* 1983.

87. **Gould, S.J. and Luria, S.E. & Singer, S.** *A View of Life.* 1981.

88. **Cracraft, J.** *Systematics, Comparative Biology and the Case Against Creationism.* 1983.

89. *Phylogeny and paleontology.* **Schaeffer, B., Hecht, M. K., and Eldredge, N.** 1972, Evolutionary Biology, Vol. 6, pp. 31-46.

90. **Gish, Duane T.** *Evolution? The fossils say no!* 3rd. San Diego : Cration-Life Pubs., 1979.

91. **Kranz, Russell.** Karl Popper's Challenge. *Creation, Social Science & Humanities Society - Quarterly Journal.* [Online] creationism.org, 1979. [Cited: 28 October 2024.] https://www.creationism.org/csshs/v02n4p20.htm.

92. **Zdelar, Tracy.** 5/13: {Part 2} Darwin & the Bible? *Hall of Fame Moms.* [Online] 16 July 2010. [Cited: 28 October 2024.] https://www.halloffamemoms.com/2010/07/513-part-2-darwin-the-bible/.

93. **Darwin, Charles.** *On the Origin of Species by Means of Natural Selection, or the Preservation of Favoured Races in the Struggle for Life.* London : Murray, 1859. p. 189. ISBN-13 : 978-1615340378.

94. **Kemper, Gary, Kemper, Hallie and Luskin, Casey.** *Discovering Intelligent Design: A Journey into the Scientific Evidence.* s.l. : Discovery Insitute, 2013.

95. **Mitchell, Tommy.** Organs of extreme perfection and complication. *Answers in Genesis.* [Online] AIG. [Cited: 28 October 2024.] https://answersingenesis.org/charles-darwin/didnt-darwin-call-the-evolution-of-the-eye-absurd/.

96. **'HotelMemory'.** Charles Darwin writing about the eye... *Redit/Christianity.* [Online] 20 November 2021. [Cited: 28 October 2024.] https://www.reddit.com/r/Christianity/comments/qy4mwm/charles_darwin_writing_about_the_evolution_of_the/.

97. **Darwin, Charles.** *On the Origin of Species by Means of Natural Selection, or the Preservation of Favoured Races in the Struggle for Life.* 6th. 1972. Chapter 6 - Organs of extreme perfection and complication..

98. **Genesis, Answers in.** The 10 Best Evidences from Science That Confirm a Young Earth. *Answers in Genesis.* [Online] 1

References

October 2021. [Cited: 321 October 2024.] https://answersingenesis.org/evidence-for-creation/10-best-evidences-young-earth/.

99. **Various.** *Geology of the mineral deposits of Australia and Papua New Guinea.* [ed.] F.E. Hughes. s.l. : Australasian Institute of Mining and Metallurgy, 1990.

100. *Geomorphic/Tectonic Control of Sediment Discharge to the Ocean: The Importance of Small Mountainous Rivers.* **Milliman, J.D. and Syvitski, J.P.M.** 5, 1992, The Journal of Geology, Vol. 100, pp. 525-544.

101. **Goodman, Richar E.** *Introduction to Rock Mechanics.* s.l. : Wiley & Sons, 1991. ISBN 13: 9780471812005.

102. **Finkelstein, Israel and Silberman, Neil Asher.** *The Bible Unearthed: Archaeology's New Vision of Ancient Israel and the Origin of Sacred Texts.* Kindle. s.l. : Free Press, 2002. pp. Location 702-711. ASIN : B00BOR8S7A.

103. **Various.** Book of Jonah - Augustine of Hippo. *Wikipedia.* [Online] [Cited: 29 October 2024.] https://en.wikipedia.org/wiki/Book_of_Jonah#Augustine_of_Hippo.

104. **Rubicondior, Rosa.** Lesson from France - The Bloody Extermination of the Cathars at Béziers - "Kill them all for the Lord knoweth them that are His!". *Rosa Rubicondior.* [Online] 23 July 2023. [Cited: 29 October 2024.] https://rosarubicondior.blogspot.com/2023/07/lesson-from-france-bloody-extermination.html.

105. —. Lesson from France - Massacre of the Cathars of Carcassonne, or How Christians Settled Theological Differences. *Rosa Rubicondior.* [Online] 22 July 2023. [Cited: 29 October 2024.] https://rosarubicondior.blogspot.com/2023/07/lesson-from-france-massacre-of-cathars.html.

106. **Lane, Nick.** *Life Ascending: The Ten Great Inventions of Evolution.* s.l. : New Scientist, 2010. ISBN-13 : 978-1861978189.

107. *Clonal genome evolution and rapid invasive spread of the marbled crayfish.* **Julian Gutekunst, Ranja Andriantsoa, Cassandra Falckenhayn, Katharina Hanna, Wolfgang Stein, Jeanne Rasamy & Frank Lyko.** s.l. : Springer Nature Lts, 5 February 2018, Nature Ecology & Evolution, Vol. 2, pp. pages567–573.

Index

Abiogenesis..141, 143, 144
Ackerman, Paul D. 151, 152, 153
Adam...203, 212, 213, 214, 215
Ailuropoda melanoleuca..129
Allele frequency.......................................53, 59, 63, 104
Allopolyploidy...64
Amarantha 107
Amaranthus tuberculatus..107
Amino acid...........................144, 145, 147, 148, 149, 150
Anoplophora glabripennis 72
Anser anser 96
Anthropic Principle, The...187
Anthropocentrism..9
Antibiotic resistance 69
Argument from ignorant incredulity, The..........148, 182
Ark 194, 195, 196, 198, 200
   Covenant, of the..195
Aspergillus nidulans 72
Attribution of Intentionality..11
Augustine of Hippo..209
Australopithecine......................128, 131, 132, 136, 138
Australopithecus 131, 132, 133, 139
   afarensis 135, 136, 138, 159
   sediba 131, 132, 133, 159
Autopolyploidy..64
Bacteria..69
Bacteriophage 69
Barriers to hybridization 58, 64, 89, 93, 98, 99
Bdelloid rotifers 73
Behe, Michael J...30, 31
Bible....4, 17, 50, 56, 75, 76, 77, 81, 151, 152, 165, 183, 193, 194, 195, 196, 204, 206, 207, 214
Bible literalism..17
Big Bang, The 141, 144
Biostratigraphy..47
Bison

bonasus 88
   caucasicus 88
  priscus 88
Bos
  primogenus 88
Branta canadensis 96
Brassica
  fruticulose 57
  napus 57
  oleracea 57
Breath of life 196
Bronze Age 17, 200, 212, 213
Cambrian 119, 120, 125, 126
Capsella 84
  grandiflora 84
  rubella 85
Carbon dating ...... *See* Carbon-14 dating
Carbon dioxide 69, 145, 146, 197
Carbon-14 dating 39, 42, 49, 50, 51, 52
Carrion crow 78
Chick, Jack 158, 170, 171
Chimerarachne yingi 130
Chimpanzee 11, 63, 77, 82, 117, 124, 128, 132, 133, 134, 135, 137, 138, 159, 171
Chloroplasts 71
Chromosome duplication 64
Chromosomes 55, 64, 65, 66, 67, 93, 99, 213
Cities of the Plain 193
Cnemidophorus
  inornatus 67
  neomexicanus 67, 68
  tigris 68
Cnemidophorus neomexicanus 67
Cognitive dissonance 13, 16, 83
Confirmation bias 13
Conspiracy 11, 15, 16
Corvus
  cornix 79, 83
  corone 79, 81, 83

Index

capellanus 80
pallescens 80
sharpii 80
corvus
   corone 79, 83
   orientalis 79
   orientalis 79
Creation Ministries 153
Creation Research, Science Education Foundation 51
Creation Week 52, 111, 120, 123
Cro Magnon Man 158, 170
Crow
   carrion 79, 89
   hooded 78, 79, 80, 89
Culex
   pipiens 114
   f molestus 113, 114
Cult 50, 54, 77, 108, 131, 144, 147, 151, 158, 194
Darwin, Charles 4, 34, 35, 55, 81, 101, 117, 124, 177, 178, 179, 180, 181, 182, 193
Darwinian gradualism 34
Darwinian Natural Selection 54, 55
Dawson, Charles 164, 165
Decay chain 40, 41
Dembski, William 30
Denisovans 63, 82, 91, 159, 169
Dinosaurs 15, 50, 51, 52, 102, 117, 118, 126, 127, 153, 155
Discovery Institute 27, 28, 32, 38
Divine punishment 11
DNA 15, 53, 55, 67, 69, 70, 72, 73, 87, 88, 91, 92, 145, 148, 170, 171
   mitochondrial 88, 105, 112
Drosophila 62
   ananassae 70
Echinocloa
   crus-galli 107
Echinocloa 107
   oryzicola 107
Ellis-van Creveld syndrome 61

Elysia chlorotica 71
Endoparasite 70
Endosymbiont 70
Endosymbiosis 71
Ensatina
   escholtzii 90
      escholtzii 90
      klauberi 90
Entacmaea medusivora 110
Environmental selectors 58, 77
Eoanthropus dawsoni 164
Escherichia coli 34
Essentalism 13
Establishment clause 30
Eukaryote 71, 73
Eve 203, 212, 213, 214, 215, 216
Evolution 11, 12, 15, 52, 53, 54, 55, 56, 58, 61, 62, 63, 66, 67, 71, 74, 77, 81, 82, 83, 85, 86, 87, 94, 96, 97, 102, 104, 105, 106, 107, 108, 109, 110, 113, 117, 118, 119, 120, 123, 124, 127, 129, 130, 137, 138, 139, 149, 150, 154, 155, 158, 162, 165, 166, 168, 169, 171, 172, 173, 174, 175, 177, 178, 193, 194
   warp speed 194
Evolutionary diversification 55, 92, 194
Evolutionary process 16, 74, 83, 129, 144, 145, 148, 150
False dichotomy 148, 150
Fine-Tuned Universe Fallacy, The 187
Fine-tuned-Universe 44
Firmament 17
Fossil record 87, 117, 119, 120, 121, 128, 154, 155, 173, 174
Founder population 60, 61, 62, 105, 111, 112, 114
Frozen Accident Theory, The 150
Futuyama, Douglas 173
Gametes 64, 65
Gene flow 101
Gene pool 35, 53, 54, 58, 59, 72, 117, 139, 147, 177
Genesis 17, 75, 76, 181, 183, 193, 195, 196, 200, 201, 202, 203, 207
Genetic bottleneck 61, 85, 194, 199

Index

Genetic code 147, 149, 150, 177
Genetic diversity 61, 62
Genetic drift 54, 55, 58, 60, 61, 62, 94, 101, 108
Genetic information 69, 78
Genocidal flood 17, 88, 193, 203
Genome 58, 59, 61, 62, 64, 67, 68, 69, 70, 71, 86, 147
Gibbons 138
Gish, Duane 178
God of the Gaps 26, 33
Gorilla 133
Grand Canyon 197
Great Ordovician Biodiversification Event 125
Ham, Ken 181, 183
Hebrew origin myths 17
Heidelberg Man 158, 159
Hesperopithecus haroldcookii 160, 161, 162
Hitchins, Christopher 26
Hominin 62, 63, 91, 92, 128, 131, 132, 133, 134, 136, 137, 138, 140, 159, 160, 169, 170
   transitional 131
Homo 91, 132, 134, 135, 136, 138, 139, 169
   africanus 159
   antecessor 82, 169
   erectus 62, 82, 91, 159, 160, 161, 168, 169
      pekinensis 168, 169
   floresensis 62
   habilis 82, 132, 159
   heidelbergensis 159, 160, 169
   neanderthalensis 63, 82, 91, 158, 159, 160, 169, 170
   sapiens 63, 82, 91, 92, 128, 134, 137, 138, 159, 160, 168, 169, 170
   sediba 132
Horizontal gene transfer 54, 68, 70, 71, 72, 73, 74
Hovind, Kent 158
Hybridization 54, 62, 63, 64, 65, 66, 67, 68, 74, 77, 83, 84, 88, 92, 97, 98, 99, 104, 107, 108, 115
Hydrogen sulphide 109, 197, 198
Intelligent design 28, 29, 30, 31, 32, 33, 34, 35, 36
Intelligent design movement 29, 31, 34, 36, 38

Intelligent designer 57
Intuitions 10
Irreducible complexity 34
Jonah 193, 207, 208, 209, 210, 211
Kind 53, 70, 74, 75, 76, 77, 78, 81, 104, 107, 113, 133, 194
Larus
  argentatus
    argentatus 89
    argenteus 89
    birulai 89
    smithsonianus, ...89
    vegae 89
  fuscus 89
    heuglini 89
    sensu stricto 89
    sensu-stricto 89
Le Gros Clarke, Professor Sir Wilfrid 166, 167
LeGros Clarke, Professor Sir Wilfred..............................163
Linnaeus
  Carl...86
Lot..193, 203, 204, 205, 206, 207
Lower Ordovician 126
Lufengpithecus, ...138
Luminescence dating 45
  optically stimulated luminescence 45
  thermoluminescence 45
Luskin, Casey...179
Lyell, Charles...180
Macro-evolution................82, 101, 102, 103, 104, 105, 124, 194
Magnetic reversals 46, 47
Mastigias
  cf. papua etpisoni 108, 110
  papua 109, 110
Methane 145, 197, 198
Micro-evolution 82, 101, 102, 103, 104, 105, 108, 123, 194
Micro-RNA 72
Middle East..17, 169, 203, 204, 215
Middle Pleistocene...160
Miller, Hugh..4, 50, 51

Index

Miller, Professor Kenneth 35
Missing link 132, 133, 136, 139, 159
Mitochondria 71
Morton's Demon 14, 15, 16, 44, 54, 131
Mutation 101
Myer, Stephen 30
Natural selection 55, 56, 57, 58, 62, 82, 85, 96, 101, 108, 122, 147, 177, 178, 181, 193
Nebraska Man 158, 160
'Nebraska Man 161, 162, 163
Nelson, Paul 30
New 170
New genetic information 69, 147
New Guinea Man 158, 170
New Guinea Man' 158
Ninevah 208, 209, 210
Nineveh 210, 211
Noah 193, 194, 196, 198, 201, 202, 203, 213
Noah's Ark 194, 213
Nucleic acid 144
O'Brien, Jonathan 154, 155, 157, 158
Oakley, Kenneth 166
O'Brien, Jonathan 153
Occam's Razor 22, 25, 26
Ockham, William of 21
Ordovician 125, 126
Original Sin 92
Osborn
   Henry Fairfield 160, 161, 162
Oxford ragwort 86
Oxygen 125, 197, 199, 200, 208
Oxyura jamaicensis 96
Palaeozoic 126
Paleomagnetic dating 46
Pan 134, 135
Parthenogenesis 67
Passer
   domesticus 79
   montanus 79

Peking Man 158, 168
Pelophylax
   kl. esculentus 65, 66
   lessonae 65
   ridibundus 65
Photosynthesis 200
Phylloscopus trochiloides 90
Piltdown Man 158, 163, 164, 167, 169
Planck Time 142
Plasmids 69, 72
Plasmodium falciparum 35
Podarcis
   melisellensis 105
   sicula 105
Polyploid 64, 65
Polyploidy 64, 65, 102
Post-zygotic barriers 93, 96, 98, 99
Potassium-Argon dating 43
Presuppositional apologetic 33
Pre-zygotic barriers 92, 93, 96, 98, 99, 115
Procambarus
   fallax 66, 67
   virginalis 66
Prokaryote 71, 146
Prosthenops 162
Quantum fluctuation 44
Quantum mechanics 11, 142, 143
Quantum zero 141
Quote mine 124, 172, 173, 174, 175, 178, 179, 180, 182
Radioactive decay rates 43, 44, 45, 192
Radioactive isotopes 40
Radio-halos 151
Radiometric geochronology 45
Reinforcement of unscientific prejudice 12
Relativity 141, 142
Retrovirus 69, 70
Ring species 82, 89, 91
RNA 70, 72, 145, 147, 148, 149
Rubidium-Strontium dating 43

Index

Sahelanthropus tchadensis 133, 134
Schweitzer, Professor Mary Higby 15
Scientific evidence 16, 17, 165, 171, 184
Second Law of Thermodynamics 68, 69, 193, 194
Selection pressures 62, 88, 95
Selection sieve 56
Senecio
   cambrensis 86
   squalidus 86
   vulgaris 86, 87
   x. baxteri 86
Sexual selection 101
Shannon Information Theory 69
Shannon, Claude 69
Sinapis alba 57
Smith Woodward, Arthur 164
Snelling, Andrew 183, 184, 185
Social Darwinism 55
Social 'Darwinism' 167, 168, 171
Soft tissue in T. rex fossil 15
Speciation 58, 62, 63, 64, 65, 66, 67, 68, 76, 82, 84, 85, 87, 88, 89, 91, 93, 94, 95, 98, 102, 103, 113, 115, 116, 175
   allopatric 57, 102
   sympatric 57
Stereochemical Hypothesis 149
Straw man 77
Taxonomy 53, 76, 77, 78, 80, 81, 105
Teleological thinking 10, 11, 15
Theory of Evolution 52, 53, 54, 55, 117, 119, 120, 121, 123, 147, 164, 167, 172, 177, 182, 193
Tower of Babel 17, 193, 200
Transitional form 117, 139
Transitional fossil 117, 119, 120, 124, 125, 128, 131, 133
Triplet code 147, 148, 149
Turdus
   merula 79
   philomelos 79
Type III secretory system 34
U-Pb dating 41, 42

Vaucheria litorea 71
Weak nuclear force 44, 192
Wedge Strategy 27, 28, 30, 32, 38
Weiner, Joseph 166
Wolbachia 70
Zircons 41, 42

Other Books by Rosa Rubicondior

(Prices correct at time of publication. Check online for current details)

The Light of Reason Series:

The Light of Reason: And Other Atheist Writings.
> Irreverent essays, thought-provoking articles and humorous items on atheism, religion, science, evolution, creationism and related issues.

	(Hardcover) ISBN-13: 979-8512173916	£13.75 (US $18.75)
	(Paperback) ISBN-10: 1516906888, ISBN-13: 978-1516906888	£9.20 (US $12.75)
	(Kindle) ASIN: B014N0IPVI	£5.50 (US $7.50)

The Light of Reason: Volume II – Atheism, Science and Evolution.
> Thought-provoking essays on the conflict between fundamentalist religion and science, and exposing the anti-science, extremist political agenda of the modern creationist industry.

	(Hardcover) ISBN-13: 979-8512191040	£13.75 (US $18.75)
	(Paperback) ISBN-10: 1517105188, ISBN-13: 978-1517105181	£9.45 (US $11.75)
	(Kindle) ASIN: B014N0IR16	£3.99 (US $5.99)

The Light of Reason: Volume III – Apologetics, Fallacies, and Other Frauds.
> Thought-provoking essays and articles on religion and atheism, dealing with religious apologetics, fallacies, miracles and other frauds

	(Hardcover) SBN-13: 979-8512526002	£13.90 (US $17.25)
	(Paperback) ISBN-10: 151710761X, ISBN-13: 978-1517107611	£7.75 (US $10.75)
	(Kindle) ASIN: B014N0IRE8	£3.50 (US $5.50)

The Light of Reason: Volume IV - The Silly Bible.
> Exposing the absurdities, contradictions and historical inaccuracies in the Bible and advancing the case for atheism and against religion. This volume, the fourth in the Light of Reason series, deals with contradictions and absurdities in the Bible.

	(Hardcover) ISBN-13: 979-8512539392	£13.75 (US $18.75)
	(Paperback) ISBN-10: 1517108209, ISBN-13: 978-1517108205	£8.22 (US $10.20)
	(Kindle) ASIN: B014N0IR8E	£3.99 (US $4.99)

The Light of Reason: And Other Atheist Writing. (all 4 volumes in one book)

Refuting Creationism

Based on the Rosa Rubicondior science and Atheism blog, this is a collection of Atheist and science articles, some short, others lengthier, exploring the interface between religion and science and which have been published over some four years.

(Kindle only) ASIN: B013DYOK32 £6.34 (US $9.95)

(Paperback) ISBN-13: 978-1521146330 £24.00 (US $30.50)

Other Books by Rosa Rubicondior

Other books on science, Atheism and theology

An Unprejudiced Mind: Atheism, Science & Reason.
Essays on science and theology from a scientific atheist perspective, exploring particularly evolution versus creationism.

(Hardcover) ISBN-13: 979-8512554685	£13.35 (US $18.75)	
(Paperback) ISBN-10: 1522925805, ISBN-13: 978-1522925804	£8.75 (US $11.75)	
(Kindle) ASIN: B019UGXPM4	£3.99 (US $5.95)	

Ten Reasons To Lose Faith: And Why You Are Better Off Without It.
Why faith is not only a fallacy and useless as a route to the truth but is actually harmful to society and to the individual. It systematically dismantles the standard religious apologetics and shows them to be bogus and deliberately constructed to mislead.

(Hardcover) ISBN-13: 979-8509108433	£18.90 (US $24.00)
(Paperback). ISBN-13:978-1530431953, ISBN–10: 1530431956	£12.60 (US $16.00)
(Kindle) ASIN: B01DGVO3JS	£6.90 (US $9.50)

What Makes You So Special? : From the Big Bang to You.
How did you come to be here, now? This book takes you from the Big Bang to the evolution of modern humans and the history of human cultures

(Hardcover) ISBN-13: 979-8509108433	£14.45 (US $18.40)
(Paperback) ISBN-13: 978-1546788294, ISBN-10: 1546788298	£10.00 (US $12.90)
(Kindle).ASIN: B071FTKXLZ	$5.69 (US $7.99)

A History of Ireland: How Religion Poisoned Everything.
From the earliest beginnings to the Northern Ireland 'Troubles' and beyond. Religion has had a major role in spreading divisions and providing excuses for subjugation and repression. Only rarely has religion played a constructive role in the development of Irish culture and political life.

(Hardcover) ISBN-13: 979-8507235032	£15.75 (US $20.00)
(Paperback): ISBN-13: 978-1724988492	£9.50 (US $12.10)
(Kindle) ASIN: B07HHHRB34	£6.10 (US $7.25)

The Internet Handbooks series

The Internet Creationists' Handbook: A Joke for the Rest of Us.
A humorous look at creationist apologetics on the Internet, showing the fallacies and dishonest tactics creationists are using to try to recruit scientifically illiterate people into their political cult.

Refuting Creationism

(Paperback),ISBN-13: 978-1721605149, ISBN-10: 1721605149   £5.78 (US $7.75)
(Kindle) ASIN: B07DZF75KD   £3.75 (US $5.00)

The Christian Apologists' Handbook: A Joke for the Rest of Us.

A humorous look at Christian apologetics on the Internet, showing the fallacies and dishonest tactics Christian fundamentalists are using to try to recruit scientifically and theologically illiterate people to their cults, often with political motives.

(Paperback) ISBN-13: 978-1721724727, ISBN–10: 1721724729   £6.25 (US $7.75)
(Kindle) ASIN: B07DYDVMW4   £3.75 (US $5.00)

The Muslim Apologists' Handbook: A Joke for the Rest of Us.

A humorous look at Muslim apologetics on the Internet, showing the fallacies and dishonest tactics Muslim fundamentalists are using to try to recruit scientifically and theologically illiterate people to their cuts, often with political motives.

(Paperback) ISBN-13: 978-1721756896, ISBN-10: 1721756892   £5.88 (US $7.75)
(Kindle) ASIN: B07DZF75KD $3.75 (US $5.00)

The Unintelligent Design Series

The Unintelligent Designer: Refuting the Intelligent Design Hoax

Showing why the superficial appearance of design in living things cannot be attributed to anything like an intelligent designer, as a counter to the politically motivated Intelligent Design movement.

(Hardcover) ISBN-13: 979-8513528463   £15.80 (US $20.00)
(Paperback) ISBN-10: 1723144215, ISBN-13: 978-1723144219   £11.20 (US $14.20)
(Kindle) ASIN B07G121BMK   £6.50 (US $7.50)

The Malevolent Designer: Why Nature's God is not Good

Showing why, if we accept, for the sake of argument, the Creationist insistence on Intelligent Design as the best explanation for biodiversity on Earth, the creator god they purport to worship could not be regarded as anything other than a malevolent evil, assiduously designing suffering into its creation as though it hates it and wants it to suffer in unimaginably horrible ways.
Illustrated by Catherine Hounslow-Webber

(Hardcover) ISBN-13: 979-8511295442   £16.70 (US $21.00)
(Paperback) SBN-13; 979-8670361729   £11.20 (US $14.15)
(Kindle) ASIN: B08L9S8F5F   £6.75 (US $8.00)

Publish under the name Bill Hounslow – Oxfordshire Childhood series.

In The Blink of an Eye: Growing Up in Rural Oxfordshire

Other Books by Rosa Rubicondior

A frank recollections of life as feral children in the small North Oxfordshire hamlet of Fawler during the 1950s and 60s, on the brink of major change as we approached the television age and the final stages in the domestication of children was about to begin. Additional material by Patricia Broome

(Hardcover) ISBN-13: 979-8511967400	£14.90 (US $19.00)
(Paperback) ISBN-10: 1545350787, ISBN-13: 978-1545350782	£8.95 (US $11.40)
(Kindle) ASIN: B06ZY8JZ92	£6.95 (US $8.95)

In The Blink of an Eye: Growing Up in Rural Oxfordshire Illustrated Editions.
Illustrated by Catherine Webber-Hounslow

(Hardcover) ISBN-13; 979-8364521361	£16.90 (US $21.60)
(Paperback): ISBN-13:979-8364503862	£10.55 (US $13.50)
(Kindle) ASIN: B0BNCQG8CC	£7.50 (US $9.10)

A Goose for Christmas: Stories from an Oxfordshire Childhood

Slightly imaginative stories, based on real events and people, of childhood adventures in the North Oxfordshire hamlet of Fawler in the 1950s during the post-war austerity, before television, when the children had only what they could get from the woods and fields around them.
Illustrated by Catherine Webber-Hounslow

(Hardcover) ISBN-13: 979-8511907482	£14.90 (US $19.00)
(Paperback) ISBN-13: 978-1981708925, ISBN-10: 1981708928	£9.35 (US $11.90)
(Kindle) ASIN: B07GFJ85P8	£6.50 (US $8.25)

www.ingramcontent.com/pod-product-compliance
Lightning Source LLC
Chambersburg PA
CBHW052310220526
45472CB00001B/56